LONDON MATHEMATICAL SOCIETY LECTURE N

Managing Editor: Professor J.W.S. Cassels, Department of Pure Math   . Mathematical Statistics,
University of Cambridge, 16 Mill Lane, Cambridge CB2 1SB, England

The books in the series listed below are available from booksellers, or, in case of difficulty,
from Cambridge University Press.

London Mathematical Society Lecture Note Series. 143

The Ergodic Theory of Discrete Groups

Peter J. Nicholls,
Professor, Department of Mathematical Sciences,
Northern Illinois University, USA

CAMBRIDGE UNIVERSITY PRESS

Cambridge

New York   Port Chester   Melbourne   Sydney

CAMBRIDGE UNIVERSITY PRESS
Cambridge, New York, Melbourne, Madrid, Cape Town, Singapore, São Paulo

Cambridge University Press
The Edinburgh Building, Cambridge CB2 8RU, UK

Published in the United States of America by Cambridge University Press, New York

www.cambridge.org
Information on this title: www.cambridge.org/9780521376747

First published 1989
Re-issued in this digitally printed version 2008

A catalogue record for this publication is available from the British Library

ISBN 978-0-521-37674-7 paperback

To Trudy

PREFACE

The interaction between ergodic theory and discrete groups has a long history and much work was done in this area by Hedlund, Hopf, Myrberg, and others over fifty years ago. During the last ten years there has been a great resurgence of interest in the area and the field is currently very active. Great advances have been made, and the theory now stands as a well developed branch of mathematical research.

The purpose of this book is two-fold. Firstly, we aim to present a connected account from first principles of the classical work in this area. Much of this material dates to the thirties, and some is difficult to locate. We will gather this material in one place with suitable explanations, simplifications, and connections drawn with the more recent body of literature. Our second aim is to present an introduction to the deep and powerful theory of measures on the limit set of a discrete group which has recently been developed by Patterson, Sullivan, and others. This circle of ideas has applications in a wide variety of problems involving discrete groups, and the notion of a measure on the limit set has emerged as one of the most powerful tools in the theory. We start from first principles and give a detailed account of the construction of the measure classes and the related conformal densities. We then consider ergodic results relative to these new measures and include a discussion of applications of these results to Hausdorff dimension of the limit set and estimates on the orbital counting function.

The book assumes a working knowledge of graduate level analysis and topology. Aside from this, every attempt has been made to keep the presentation self contained. Many of the results we give are to be found in the literature and we have attempted to provide the correct attribution. The proofs that are given here have in some cases been taken directly from the source but for the most part have been constructed by combining the ideas of more than one author.

This book grew out of a fascination for the marvelous work of Dennis Sullivan and S.J. Patterson. Their contributions in recent years to the theory of discrete groups have been astounding, and the theory covered in this book represents just one area of their influence. A major proportion of the results we present are due to Sullivan, although this has not always been made explicit.

With the major purpose of providing the introduction and background necessary for an understanding of, and an appreciation for, measures on the limit set of a discrete group, there are many parts of the subject which are not mentioned — indeed to cover them all would occupy several volumes such as this. In particular we have not touched on symbolic dynamics, nor on the connection between our conformal measures and Gibbs measures recently exploited in the beautiful work of Rees. The survey article of Patterson [Patterson, 1987] should be consulted for further information on these topics and for a good bibliography.

There are many individuals who have helped in the preparation of this book. My friend and teacher Alan Beardon has been a constant source of encouragement and support, and has provided crucial assistance at several stages of the project. My colleague Peter Waterman has read the entire manuscript with great care and provided many valuable insights. Lars Ahlfors has graciously permitted me to quote extensively from his beautifully written Minnesota lecture notes. James Norris has been a very supportive Dean, providing me both with facilities and with time for the completion of the project. Thanks are also due to Sara Clayton for help with the preparation of the manuscript. I would like also to express gratitude to the National Science Foundation for their support during the early stages of the project. Of course, I owe a major debt of gratitude to my family who have put up with my constant preoccupation with "the book" over a long period of time.

Peter Nicholls
DeKalb Illinois
March 1989

CONTENTS

Contents

CHAPTER 1

Preliminaries

1.1 Area

One of our main concerns in this book is with the measure of certain subsets of the unit sphere. Accordingly, in this section, we introduce Lebesgue surface area on the unit sphere in R^n.

Throughout we denote by B the unit ball in R^n and by S the unit sphere. Thus, denoting points of R^n by $x = (x_1, x_2, ..., x_n)$ and writing $|x| = \left(\sum x_i^2\right)^{1/2}$ we have

$$B = \{x : |x| < 1\} \quad \text{and} \quad S = \{x : |x| = 1\}.$$

Note that we will throughout write vectors as rows, but in operations involving matrices, they will be interpreted as columns. Given a point $(x_1, x_2, ..., x_n)$ in R^n we introduce polar coordinates as follows. Set $r^2 = \sum x_i^2$ and, for j satisfying $1 \le j < n$, define θ_j to be the angle between the jth coordinate vector and the vector $(0, 0, ..., 0, x_j, x_{j+1}, ..., x_n)$. Thus

$$\theta_j = \arccos \ [x_j/(x_j^2 + ... + x_n^2)^{1/2}] \quad \theta_j \in [0, \pi] \quad \text{for} \ \ 1 \le j < n-1 ,$$

$$\theta_{n-1} = \begin{cases} \arccos \ [x_{n-1}/(x_{n-1}^2 + x_n^2)^{1/2}] \ \text{if} \ x_n \ge 0 \\ 2\pi - \arccos \ [x_{n-1}/(x_{n-1}^2 + x_n^2)^{1/2}] \ \text{if} \ x_n < 0 \end{cases} \quad \theta_{n-1} \in [0, 2\pi) .$$

Proceeding inductively, we can show that

$$x_1 = r\cos\theta_1$$
$$x_j = r\sin\theta_1\sin\theta_2\cdots\sin\theta_{j-1}\cos\theta_j \quad \text{for} \quad 1 < j < n .$$
$$x_n = r\sin\theta_1\sin\theta_2\cdots\sin\theta_{n-1}$$

The Jacobian of the transform $(r,\theta_1,\theta_2,\ldots,\theta_{n-1}) \to (x_1,x_2,\ldots,x_n)$ is calculated inductively to be

$$r^{n-1}(\sin\theta_1)^{n-2}(\sin\theta_2)^{n-3}\ldots(\sin\theta_{n-2}).$$

Thus the volume element in polar coordinates is

$$dV = r^{n-1}(\sin\theta_1)^{n-2}(\sin\theta_2)^{n-3}\cdots(\sin\theta_{n-2})d\theta_1 d\theta_2 \cdots d\theta_{n-1}dr$$

and the surface area element on the unit sphere S is

$$dw = (\sin\theta_1)^{n-2}(\sin\theta_2)^{n-3}\cdots(\sin\theta_{n-2})d\theta_1 d\theta_2 \cdots d\theta_{n-1}.$$

We shall have occasion to measure the surface area of a subset of S interior to a ball and the following lemma is useful.

Lemma 1.1.1   For $\eta \in S$ and $\lambda > 0$, set $A = \{x \in S : |x - \eta| < \lambda\}$. Then

$$w(A) = M \int_0^{\mu}(\sin\theta)^{n-2}d\theta$$

where $\mu = \arccos(1 - \lambda^2/2)$, and M is an absolute constant.

Proof. We may as well assume that $\eta = (1,0,0,\ldots,0)$ and then, for $x \in S$, $|\eta - x|^2 = 2(1 - x_1)$ where $x = (x_1,x_2,\ldots,x_n)$. From this it follows that $A = \{x \in S : 1 - \lambda^2/2 \le x_1 \le 1\}$. Recall however that $x_1 = \cos\theta_1$ and the lemma follows when we integrate the surface area element over A and set

$$M = \int_0^{2\pi}\int_0^{\pi}\ldots\int_0^{\pi}(\sin\theta_2)^{n-3}(\sin\theta_3)^{n-4}\cdots(\sin\theta_{n-2})d\theta_2 \cdots d\theta_{n-1}$$

which is an absolute constant — namely, the (n-2)-dimensional Lebesgue measure of the unit sphere in R^{n-1}. $\square$

We shall be much concerned in later chapters with the Hausdorff measure and dimension of various subsets of the unit sphere, and we briefly review the definitions. Suppose E is a Borel set in R^n and $\alpha > 0$ is given, if we denote by $\Delta(x,c)$ the ball centered at x of Euclidean radius c, then we define for $\epsilon > 0$

$$\Lambda_\alpha^\epsilon(E) = \inf\{\sum_{j=1}^{\infty} c_j^\alpha : E \subset \bigcup \Delta(x_j,c_j) ; c_j \le \epsilon\}.$$

This clearly decreases as ϵ increases and the (possibly infinite) limit

$$\Lambda_\alpha(E) = \lim_{\epsilon \to 0} \Lambda_\alpha^\epsilon(E)$$

exists. This quantity is called the α-dimensional Hausdorff measure of E. The Hausdorff dimension, $d(E)$, of a Borel set E is defined by

$$d(E) = \inf\{\alpha : \Lambda_\alpha(E) = 0\} = \sup\{\alpha : \Lambda_\alpha(E) = +\infty\}. \qquad (1.1.1)$$

A consequence of this definition is that if $0 < \Lambda_\alpha(E) < +\infty$ then $d(E) = \alpha$.

1.2 The Hyperbolic Space

The unit ball B of R^n is a model for n-dimensional hyperbolic space and supports a metric ρ derived from the differential

$$d\rho = \frac{2|dx|}{1 - |x|^2}.$$

Lines in the space are arcs of circles orthogonal to the unit sphere S and are geodesics for the metric ρ. Angle is Euclidean angle.

An alternative model of n-dimensional hyperbolic space is the upper half space H of R^n

$$H = \{x = (x_1, x_2, ..., x_n) : x_n > 0\}$$

together with the metric ρ derived from the differential

$$d\rho = \frac{|dx|}{x_n}.$$

Lines in this space are arcs of circles orthogonal to the plane $\{x : x_n = 0\}$ and are geodesics for the metric ρ.

Note that we use ρ for the metric in both the ball and the upper half space model — no confusion should arise. We shall be working almost entirely in the ball but will from time to time be using the upper half space model, each has its own particular advantages.

There is a wealth of information on hyperbolic geometry to be found in [Beardon, 1983] and in [Ahlfors, 1981, Chapter 3] — we will quote extensively from these sources. The formula for the hyperbolic distance between a point and a line is to be found in [Beardon, 1983, p.162] but not in the form best suited for our purposes. We state the result now, although we will not be in a position to prove it until the next section.

Theorem 1.2.1   Suppose $a \in B$ and $\eta, \xi \in S$, $\eta \neq \xi$. Let s be the hyperbolic distance from a to the geodesic joining ξ and η then

$$\cosh s = \frac{2|a - \xi||a - \eta|}{|\xi - \eta|(1 - |a|^2)}.$$

For later purposes we introduce the notion of "shadows". Suppose $a \in B$ and

$\delta > 0$ are given, and let $\Delta(a,\delta)$ denote the hyperbolic ball of center a and radius δ. The set $b(0:a,\delta)$ is the projection from the origin onto S of the ball $\Delta(a,\delta)$ and so every point of $b(0:a,\delta)$ lies in the shadow of the ball. A point $\xi \in S$ clearly belongs to $b(0:a,\delta)$ if and only if the radius to ξ passes within a hyperbolic distance δ of a. From theorem 1.2.1 this is equivalent to

$$|a - \xi||a + \xi| < (1 - |a|^2)\cosh \delta \quad \text{and} \quad |a + \xi| > |a - \xi|. \quad (1.2.1)$$

It will be useful to compute the w-measure of the shadow $b(0:a,\delta)$ and we have the following.

Theorem 1.2.2   For $\delta > 0$

$$w(b(0:a,\delta)) \sim \frac{M(\cosh^2\delta - 1)^{(n-1)/2}}{n-1}(1 - |a|)^{n-1}$$

uniformly as $|a| \to 1$ in B, where M is the constant of lemma 1.1.1.

Proof. Suppose $a \in B$ and $\xi \in S$ then

$$|a - \xi|^2 = 1 + |a|^2 - 2a.\xi,$$
$$|a + \xi|^2 = 1 + |a|^2 + 2a.\xi,$$

and

$$\left|\xi - \frac{a}{|a|}\right|^2 = 2 - \frac{2a.\xi}{|a|}.$$

We use these equations and inequality (1.2.1) to note that $\xi \in b(0:a,\delta)$ if and only if

$$(1 + |a|^2)^2 - 4(a.\xi)^2 < (1 - |a|^2)^2 \cosh^2\delta$$

in other words,

$$(1 + |a|^2)^2 - (2 - |\xi - a/|a||^2)^2 |a|^2 < (1 - |a|^2)^2\cosh^2\delta.$$

This reduces to

$$\left|\xi - \frac{a}{|a|}\right| < \frac{[2|a| - ((1 + |a|^2)^2 - (1 - |a|^2)^2 \cosh^2\delta)^{1/2}]^{1/2}}{|a|^{1/2}} = \lambda,$$

say. A routine calculation shows that

$$\lambda \sim (1 - |a|)(\cosh^2\delta - 1)^{1/2} \quad as \quad |a| \to 1. \quad (1.2.2)$$

Using lemma 1.1.1 we see that

$$w(S(a,\delta)) = M \int_0^\mu (\sin\,\theta)^{n-2} d\theta \qquad (1.2.3)$$

where $\mu = arccos\,(1 - \lambda^2/2)$. Clearly $\mu \sim (1 - |a|)(cosh^2\delta - 1)^{1/2}$ as $|a| \to 1$ (from (1.2.2)) and, for θ between 0 and μ, we approximate $\sin\theta$ by θ. The theorem now follows from (1.2.3). □

In order to measure the size of various subsets of the unit sphere, the following definition is useful. If $a \in B$ and $k, \alpha > 0$ we define

$$I(a:k,\alpha) = \left\{ x \in S : \left| x - \frac{a}{|a|} \right| < k(1 - |a|)^\alpha \right\}. \qquad (1.2.4)$$

We next consider cones at a point $\xi \in S$. If $x \in B, \xi \in S$ and λ satisfies $0 < \lambda < \pi/2$ then we say x belongs to the cone at ξ of opening λ if the angle between the vectors ξ and $\xi - x$ is at most λ and, further, $|x - \xi| < 2\cos\lambda$. The cosine of the angle between ξ and $\xi - x$ is calculated to be

$$\frac{\xi.(\xi - x)}{|\xi||\xi - x|} = \frac{2 - 2\xi.x}{2|\xi - x|} = \frac{(\xi - x).(\xi - x) + 1 - |x|^2}{2|\xi - x|} = \frac{|\xi - x|^2 + 1 - |x|^2}{2|\xi - x|}$$

and we have proved the following.

Lemma 1.2.3   If $x \in B, \xi \in S$ and λ satisfies $0 < \lambda < \pi/2$ then x belongs to the cone at ξ of opening λ if and only if $|x - \xi| < 2\cos\lambda$ and

$$\frac{|\xi - x|^2 + 1 - |x|^2}{2|\xi - x|} > \cos\lambda.$$

Theorem 1.2.4   Suppose $\xi \in S$ and $\{x_n\}$ is a sequence of points of B with $|x_n| \to 1$ as $n \to \infty$. The following are equivalent.

1. There exists $a > 0$ such that, for n large enough, x_n lies in the cone of opening a at ξ.

2. There exists $b > 1$ such that, for n large enough,

    $$|x_n - \xi| < b(1 - |x_n|).$$

3. There exists $c > 0$ such that, for n large enough,

    $$\xi \in I(x_n:c,1).$$

4. There exists $d > 0$ such that if l is any geodesic ending at ξ then, for n large enough, $\rho(x_n,l) < d$.

Proof. If (1) is true we note that, for n large enough,

$$\frac{|\xi - x_n|^2 + 1 - |x_n|^2}{2|\xi - x_n|} > \cos a$$

from lemma 1.2.3. Since $|x_n| \to 1$ we see that, given $\epsilon > 0$, for n large enough,

$$|\xi - x_n| < \frac{(1 - |x_n|)}{(\cos a - \epsilon)}.$$

Thus (1) implies (2). Now suppose (2) is true and we note that

$$\left|\xi - \frac{x_n}{|x_n|}\right|^2 = \frac{|\xi - x_n|^2 - (1 - |x_n|)^2}{|x_n|}.$$

Therefore

$$\left|\xi - \frac{x_n}{|x_n|}\right| < \frac{(b^2 - 1)^{1/2}(1 - |x_n|)}{|x_n|^{1/2}}.$$

We may take $c = (b^2 - 1)^{1/2} + \epsilon$ and (3) follows. Assuming (3) we let l be a geodesic ending at ξ and suppose that η is the other end point of l. From theorem 1.2.1 we see that

$$\cosh \rho(x_n, l) = \frac{2|x_n - \xi||x_n - \eta|}{|\xi - \eta|(1 - |x_n|^2)}$$

which is asymptotic (as $n \to \infty$) to $\frac{|x_n - \xi|}{(1 - |x_n|)}$. However, from (3)

$$\frac{|\xi - x_n|}{1 - |x_n|} < (1 + c^2)^{1/2}$$

for n large enough and (4) is true. Finally, we suppose (4) is true and note from theorem 1.2.1 that, for n large enough,

$$\frac{2|x_n - \xi||x_n - \eta|}{|\xi - \eta|(1 - |x_n|^2)} < \cosh d.$$

Thus, from our remarks above, if $\epsilon > 0$ is given then $\frac{|x_n - \xi|}{(1 - |x_n|)} < \cosh d + \epsilon$.

For n large enough,

$$\frac{|\xi - x_n|^2 + 1 - |x_n|^2}{2|\xi - x_n|} > \frac{1}{\cosh d + \epsilon} - \epsilon$$

and (1) follows from lemma 1.2.3. This completes the proof of the theorem. $\square$

    It should be noted, from our working above, that the constants a, b, c, d are related by

$$b \approx \frac{1}{\cos a} \approx \cosh d \approx (1 + c^2)^{1/2}$$

In other words, (1) implies (2) with $b = \frac{1}{\cos a} + \epsilon$ for any $\epsilon > 0$ and (2) implies
(1) with $a = \arccos(1/b) + \epsilon$ for any $\epsilon > 0$. Similar remarks hold for the
relations between b, c and d.

We next consider horospheres. A **horosphere** at $\xi \in S$ is a sphere in R^n
which is internally tangent to the unit sphere S at ξ. A **horoball** is the interior
of a horosphere.

Theorem 1.2.5   If $\xi \in S$, $x \in B$ and $0 < k < 1$ then x is on the horosphere at
ξ of Euclidean radius k if and only if

$$(1 - |x|^2)|x - \xi|^{-2} = \frac{1 - k}{k}.$$

The point x is in the horoball at ξ of radius k if and only if

$$(1 - |x|^2)|x - \xi|^{-2} > \frac{1 - k}{k}.$$

Proof. Suppose $x \in B$ with $(1 - |x|^2)|x - \xi|^{-2} = \frac{1 - k}{k}$ then

$$1 - |x|^2 = \frac{1 - k}{k}(x - \xi).(x - \xi) = \frac{1 - k}{k}(|x|^2 + 1 - 2x.\xi)$$

and so

$$2x.\xi = \frac{(|x|^2 + 1 - 2k)}{(1 - k)}. \tag{1.2.5}$$

Now the square of the distance of x from the center of the horosphere is

$$|x - (1 - k)\xi|^2 = |x|^2 + (1 - k)^2 - 2(1 - k)x.\xi$$

$$= |x|^2 + (1 - k)^2 - (|x|^2 + 1 - 2k)$$

$$= k^2$$

where we substituted for $x.\xi$ from (1.2.5) above. Thus x lies on the horosphere.
Our argument is clearly reversible and we have the if and only if condition. The
statement concerning the horoball is an easy modification. $\square$

1.3 Moebius Transforms

In this section we consider Moebius transforms acting in R^n, derive several of
their properties, and show that those preserving the ball are isometries of

hyperbolic space. For a full and detailed account, the reader is referred to [Beardon, 1983, Chapter 3] or [Ahlfors, 1981, Chapter 2]. Much of what we do in this section is in fact a slightly compressed version of Ahlfors' account. We start by defining a **similarity** as a map $R^n \to R^n$ given by

$$x \;\to\; m\,x + b$$

where $b \in R^n$ and m is a **conformal** matrix (i.e., a positive constant multiple of an orthogonal matrix). Reflection in the unit sphere is given by

$$x \;\to\; x^* \;=\; J(x) \;=\; \frac{x}{|x|^2}$$

and we define the full Moebius group as the group generated by J and all the similarities.

The derivative of a self map of R^n is the Jacobian matrix, and we will use the prime notation. Observe that the derivative of a similarity $\gamma(x) = m\,x + b$ is the constant matrix m. In order to write down the derivative of J we introduce the matrix $Q(x)$ by

$$Q(x)_{ij} \;=\; \frac{x_i\,x_j}{|x|^2}$$

and leave it as an exercise to check that, for $x \neq 0$,

$$J'(x) \;=\; \frac{1}{|x|^2}\,[I - 2Q(x)]. \tag{1.3.1}$$

Since $Q^2 = Q$ we have $[I - 2Q]^2 = I$ and it follows that $I - 2Q$ is an orthogonal matrix. For each $x \neq 0$, $J'(x)$ is a conformal matrix.

Use of the chain rule shows that $\gamma'(x)$ is a conformal matrix for any Moebius γ — in other words, Moebius transforms are **conformal**. For any Moebius γ we denote by $|\gamma'(x)|$ the positive number such that $\gamma'(x)/|\gamma'(x)|$ is orthogonal. Thus $|\gamma'(x)|$ is the linear change of scale at x, the same in all directions.

The following equation, which will be fundamental to our work, follows from the chain rule and application of (1.3.1).

$$|\gamma(x) - \gamma(y)| \;=\; |\gamma'(x)|^{1/2}\,|\gamma'(y)|^{1/2}\,|x - y|. \tag{1.3.2}$$

Application of (1.3.2) proves the invariance of the absolute cross ratio

$$|a,b,c,d| \;=\; \frac{|a-c|}{|a-d|} \cdot \frac{|b-d|}{|b-c|}$$

in the sense that

$$|\gamma(a),\gamma(b),\gamma(c),\gamma(d)| \;=\; |a,b,c,d|. \tag{1.3.3}$$

We denote by $GM(B)$ the subgroup of the full Moebius group which leaves the unit ball invariant, and prove the following lemma.

Lemma 1.3.1   If $\gamma \in GM(B)$ and $\gamma(0)=0$ then γ is a rotation. In other words, $\gamma(x) = k\,x$ where k is an orthogonal matrix.

Proof. (Ahlfors, 1981 p.21) Let us suppose first that $\gamma(\infty)=\infty$. Since

$$|\gamma(x),\gamma(y),0,\infty| \;=\; |x,y,0,\infty|$$

we deduce that $|\gamma(x)|/|x| = \lambda$, a constant. The equation

$$|\gamma(x),0,\gamma(y),\infty| \;=\; |x,0,y,\infty|$$

yields $|\gamma(x)-\gamma(y)|^2 = \lambda^2|x-y|^2$. It follows that for any x $\gamma(x)=\lambda^2 x$ and so

$$|\gamma(x+y)-\gamma(x)-\gamma(y)|^2 \;=\; \lambda^2|(x+y)-x-y|^2 \;=\; 0.$$

Thus $\gamma(x+y)=\gamma(x)+\gamma(y)$. From this we deduce that γ' is constant, and $\gamma(x)=m\,x$ with a constant conformal matrix m. Since $|mx|=1$ for $|x|=1$ we have that m is orthogonal as required.

If we now suppose that $\gamma^{-1}(\infty)=b \neq \infty$ then the Moebius transform $\gamma((x-b^*)^*+b)$ fixes 0 and ∞ and so, by our working above, we have

$$\gamma(x) \;=\; m\,((x-b)^*+b^*)$$

for a constant conformal matrix m. But γ preserves the unit ball and so $|(x-b)^*+b^*|$ is a constant for $|x|=1$. But

$$|(x-b)^*+b^*| \;=\; \frac{|x|}{|x-b||b|}$$

and so $|x-b|$ is constant on the unit sphere — this is impossible since $b\neq 0$, and the contradiction completes the proof of the lemma. $\square$

With this result in hand, we determine the form of the most general Moebius transform preserving B. From now on, indeed for the rest of the book, we confine attention to the orientation preserving Moebius transforms preserving B. These are the transforms containing an even number of factors J and sense preserving similarities. Thus, from this point on, $M(B)$ will be used to denote the group of orientation preserving Moebius transforms preserving B.

Given $a \in B$, $a \neq 0$, let S_a be the sphere centered at a^* and of radius $(1-|a|^2)^{1/2}/|a|$ — this sphere intersects the unit sphere orthogonally. Let σ_a denote reflection in S_a so that

$$\sigma_a(x) \;=\; a^* + (\,|\,a^*\,|^2 - 1)(x - a^*)^*.$$

We write R_a for the reflection in the plane through the origin perpendicular to a and define the canonical map

$$T_a(x) \;=\; R_a \circ \sigma_a(x).$$

As an immediate consequence of lemma 1.3.1, we note that the most general Moebius transform preserving the ball and mapping a to 0 can be written as T_a followed by a rotation.

We need to derive explicit formulae for $T_a(x)$ and note first that the reflection R_a amounts to multiplication by the matrix $I - 2Q(a)$. This leads immediately to

$$T_a(x) \;=\; -a^* + (\,|\,a^*\,|^2 - 1)\,[I - 2Q(a)]\,(x - a^*)^*. \qquad (1.3.4)$$

To proceed further we need a lemma.

Lemma 1.3.2   For any x, y

$$(x - y^*)^* + y \;=\; |\,y\,|^2\,[I - 2Q(y)]\,(x^* - y)^*.$$

Proof. (Ahlfors, 1981 p.22) Consider two transforms

$$A(x) = (x - y^*)^* \quad \text{and} \quad B(x) = |\,y\,|^2\,[I - 2Q(y)]\,(x^* - y)^* - y$$

and note that $0, \infty$ are fixed points of AB^{-1}. As in the proof of lemma 1.3.1, we see that $(AB^{-1})'$ is constant. Now note that

$$(AB^{-1})'(x) = A'(B^{-1}(x)) \cdot (B'(B^{-1}(x)))^{-1} = (A'(B')^{-1}) \circ B^{-1}(x)$$

which is constant. Thus, for some λ, $A'(x) = \lambda B'(x)$. Differentiation yields

$$A'(x) \;=\; \frac{I - 2Q(x - y^*)}{|\,x - y^*\,|^2}$$

$$B'(x) \;=\; \frac{|\,y\,|^2\,(I - 2Q(y))\,(I - 2Q(x^* - y))\,(I - 2Q(x))}{|\,x^* - y\,|^2\,|\,x\,|^2}.$$

So for $x = y$ we obtain

$$A'(y) = B'(y) = \frac{(I - 2Q(y))\,|\,y\,|^2}{(1 - |\,y\,|^2)^2}.$$

It follows that $A'(x) = B'(x)$ for all x and, since $A(0) = B(0) = -y$, we have $A(x) = B(x)$ and this completes the proof. $\square$

Note from the formulae for A' and B' that

$$(I - 2Q(y))(I - 2Q(x - y^*)) \;=\; (I - 2Q(x^* - y))(I - 2Q(x)). \quad (1.3.5)$$

Replace y by a in lemma 1.3.2 and multiply by $I - 2Q(a)$ to obtain

$$(I - 2Q(a))(x - a^*)^* \;=\; a + |a|^2(x^* - a)^*$$

and, using this in (1.3.4), we have

$$T_a(x) \;=\; -a^* + a(|a^*|^2 - 1) + (1 - |a|^2)(x^* - a)^*.$$

This reduces to

$$T_a(x) \;=\; -a + (1 - |a|^2)(x^* - a)^*. \quad (1.3.6)$$

This formula is easy to differentiate and yields

$$T_a'(x) \;=\; \frac{(1 - |a|^2)[I - 2Q(x^* - a)][I - 2Q(x)]}{|x^* - a|^2 |x|^2}.$$

Introducing the notation

$$\Delta(x,a) \;=\; [I - 2Q(x^* - a)][I - 2Q(x)] \quad (1.3.7)$$

we have

$$T_a'(x) \;=\; \frac{1 - |a|^2}{|x^* - a|^2 |x|^2}\,\Delta(x,a). \quad (1.3.8)$$

Recalling that $I - 2Q(a)$ is orthogonal for any a, we see that $\Delta(x,a)$ is orthogonal and so

$$|T_a'(x)| \;=\; \frac{1 - |a|^2}{|x^* - a|^2 |x|^2}. \quad (1.3.9)$$

Our next task is to compute $|T_a(x)|$. For this we use the difference formula (1.3.2) to obtain

$$|T_a(x)| = |T_a(x) - T_a(a)| = |T_a'(x)|^{1/2} |T_a'(a)|^{1/2} |x - a|$$

and so $|T_a(x)| = |x - a|\,|x|^{-1}\,|x^* - a|^{-1}$. An easy computation leads to

$$1 - |T_a(x)|^2 \;=\; \frac{(1 - |x|^2)(1 - |a|^2)}{|x|^2 |x^* - a|^2}$$

and, combining this with (1.3.9),

$$\frac{|T_a'(x)|}{1 - |T_a(x)|^2} \;=\; \frac{1}{1 - |x|^2}.$$

Thus we have proved the following.

Theorem 1.3.3   The metric ρ is invariant under all Moebius transforms preserving B. Specifically, for any $x \in B$ and any Moebius γ,

$$1 - |\gamma(x)|^2 = |\gamma'(x)|(1 - |x|^2).$$

As a consequence we derive the following.

Theorem 1.3.4   If γ is a Moebius transform preserving B and $\xi \in S$ then

$$|(\gamma^{-1})'(\xi)| = (1 - |\gamma(0)|^2)|\xi - \gamma(0)|^{-2}.$$

Proof. From the chain rule, $(\gamma^{-1})'(\gamma(\xi)) = (\gamma'(\xi))^{-1}$ which, when ξ is replaced by $\gamma^{-1}(\xi)$ yields

$$(\gamma^{-1})'(\xi) = (\gamma'(\gamma^{-1}(\xi))^{-1}. \qquad (1.3.10)$$

Now consider (1.3.2) with $x = \gamma^{-1}(\xi)$ and $y = 0$, and we have

$$|\xi - \gamma(0)|^2 = |\gamma'(\gamma^{-1}(\xi))||\gamma'(0)|.$$

From theorem 1.3.3 we see that $1 - |\gamma(0)|^2 = |\gamma'(0)|$ and so

$$|\xi - \gamma(0)|^2 = |\gamma'(\gamma^{-1}(\xi))|(1 - |\gamma(0)|^2)$$

which, in view of (1.3.10), is the required result. $\square$

It will be important to us later to understand how the matrix $\Delta(x,y)$ behaves under the action of a Moebius transform preserving B.

Lemma 1.3.5   If γ is a Moebius transform preserving B and if $x, y \in B$ then

$$\Delta(\gamma(x), \gamma(y)) \frac{\gamma'(x)}{|\gamma'(x)|} = \frac{\gamma'(y)}{|\gamma'(y)|} \Delta(x, y).$$

Proof. (Ahlfors, 1981 p.29) We first show that

$$T_{\gamma(y)}(\gamma(x)) = \frac{\gamma'(y)}{|\gamma'(y)|} T_y(x). \qquad (1.3.11)$$

Let $L(x)$ and $R(x)$ denote the left and right hand sides of (1.3.11). Clearly $L(y) = R(y) = 0$ and so $LR^{-1}(0) = 0$. However,

$$L'(y) = \frac{\gamma'(y)}{1 - |\gamma(y)|^2} = R'(y)$$

the left equality being obvious and the right following from theorem 1.3.3. Thus $(LR^{-1})'(0) = L'(y)(R'(y))^{-1} = I$ and so $LR^{-1} = I$ — this proves (1.3.11). Differentiation of (1.3.11) yields

$$T_{\gamma(y)}{}'\,(\gamma(x))\,\gamma'\,(x)\ =\ \frac{\gamma'\,(y)}{|\gamma'\,(y)|}\,T_y{}'\,(x)$$

and the lemma follows directly from (1.3.8) when we compare the orthogonal matrix parts of both sides. □

Our next task is a classification of Moebius transforms. This classification is based upon fixed points — note that any Moebius transform γ preserving B also preserves $\overline{B}$ and thus, by the Brouwer fixed point theorem, fixes some point of $\overline{B}$. A transform with precisely one fixed point which lies on S is said to be **parabolic**, a transform with precisely two fixed points which lie on S is said to be **loxodromic**, any other transform is **elliptic**. We need to consider parabolic transforms in some further detail.

It will be convenient to start our discussion in the upper half space model

$$H\ =\ \{(x_1, x_2, ..., x_n)\in R^n : x_n > 0\}$$

where we assume that the parabolic transform under consideration preserves H and fixes ∞. It is well known (see [Beardon, 1983, p.40] for example) that any transform ϕ preserving H and fixing ∞ may be written in the form

$$\phi(x)\ =\ r\,Ax + x_0$$

where $r > 0$, $x_0 = (x_0^1, x_0^2, \ldots, x_0^{n-1}, 0)\in R^n$, and A is a matrix of the form

$$A\ =\ \begin{bmatrix} B & 0 \\ 0 & 1 \end{bmatrix}$$

where B is an $(n-1)\times(n-1)$ orthogonal matrix. Since A is orthogonal, we see that $(rA - I)$ is invertible for $r \neq 1$ and so a fixed point of ϕ (other than ∞) exists in this case. Since our transform is parabolic, we must have $r = 1$. Given the form of A and x_0 the following result is evident.

Theorem 1.3.6   If ϕ is a parabolic transform preserving H and fixing ∞ then, for any $k > 0$, the $(n-1)$-dimensional hyperplane $\{x_n = k\}$ is preserved by ϕ.

We may assume (by a conjugation if necessary) that the matrix A discussed above is of the form

$$\begin{pmatrix} A_1 & & & & \\ & A_2 & & & \\ & & \cdot & & \\ & & & \cdot & \\ & & & & \cdot & \\ & & & & A_r & \\ & & & & & I_s \end{pmatrix}$$

— see [Beardon, 1983, p.25] for example — where r is a non-negative integer, s is a positive integer, I_s is the $s \times s$ identity matrix, and

$$A_k = \begin{pmatrix} \cos\theta_k & -\sin\theta_k \\ \sin\theta_k & \cos\theta_k \end{pmatrix}$$

We assume in this formulation that $\cos\theta_j \neq 1$, $j = 1,2,...,r$ and that $s \geq 2$ (otherwise ϕ has a fixed point in H). With these assumptions, and writing $\phi(x) = Ax + x_0$, we set $\alpha = (x_0^1,...,x_0^{2r},0,0,...,0)$ and $\beta = (0,0,...,0,x_0^{2r+1},...,x_0^{n-1},0)$. If we now define a new transform R by $R(x) = xA + \alpha$ we note that R has fixed points in H and is thus an elliptic transform. Now define T by $T(x) = x + \beta$ and note that $RoT = ToR = \phi$. We have proved the following.

Theorem 1.3.7   If ϕ is a parabolic transform preserving H and fixing ∞ then there exists an elliptic transform R and a pure translation T such that

$$\phi = ToR = RoT$$

further, R and T are unique.

Thus to any parabolic transform fixing ∞ we may associate a unique translation vector β as given by theorem 1.3.7. Note further that ϕ acts as a pure translation on the plane $(0,0,...,0,x^{2r+1},x^{2r+2},...,x^{n-1},0)$. We call this plane the **translation plane** of ϕ.

We come next to the proof of theorem 1.2.1. Consider first the special case where the geodesic joining ξ to η is a diameter of the ball and the geodesic from a which is orthogonal to this diameter is another diameter. In this case the distance s from the geodesic to a is just $\rho(0,a)$. But $\rho(0,a) = \log\left((1 + |a|)/(1 - |a|)\right)$ and so

$$\cosh[\rho(0,a)] = \frac{1 + |a|^2}{1 - |a|^2} = \frac{|a - \xi||a - \eta|}{1 - |a|^2}$$

which is the required result.

We now proceed to the general case and let x be the point on the geodesic joining ξ to η which is closest to a. Let γ be a Moebius transform preserving B with $\gamma(x) = 0$. Since the hyperbolic distance is invariant under Moebius transforms we have, using the special case already proved,

$$\cosh s = \frac{|\gamma(a) - \gamma(\xi)||\gamma(a) - \gamma(\eta)|}{1 - |\gamma(a)|^2}$$

$$= \frac{|\gamma'(a)||\gamma'(\xi)|^{1/2}|\gamma'(\eta)|^{1/2}|a - \xi||a - \eta|}{|\gamma'(a)|(1 - |a|^2)}$$

$$= \frac{|\gamma(\xi) - \gamma(\eta)||a - \xi||a - \eta|}{|\xi - \eta|(1 - |a|^2)}$$

$$= \frac{2|a - \xi||a - \eta|}{|\xi - \eta|(1 - |a|^2)}$$

where we have used (1.3.2). This is the required result. $\square$

From time to time we will need to relate Moebius transforms preserving H to those preserving B. Consider therefore the transform V which maps B onto H and is defined by

$$V(y_1, y_2, \ldots, y_n) = (x_1, x_2, \ldots, x_n)$$

where

$$x_i = \frac{2 y_i}{|y - e_n|^2} \qquad i = 1, 2, 3, \ldots, n-1$$

$$x_n = \frac{1 - |y|^2}{|y - e_n|^2}$$

with e_n denoting the vector $(0, 0, 0, \ldots, 1)$. The map V is a Moebius transform of R^n [Ahlfors, 1981, p.35] and will be used extensively. Note in particular that γ is a Moebius transform preserving H if and only if $V^{-1}\gamma V$ is a Moebius transform preserving B.

1.4 Discrete Groups

As in the previous section, we denote by $M(B)$ the full group of Moebius transforms preserving B. Recall that the most general element of $M(B)$ may be written as a canonical transform T_a followed by a rotation. It is clear then that $M(B)$ may be topologized by $O(n) \times B$. A subgroup Γ of $M(B)$ is **discrete** if the identity has a neighborhood whose intersection with Γ reduces to the identity. We comment in passing that a discrete group is necessarily countable.

For $a \in B$ we define the orbit of a, $\Gamma(a)$, to be the set $\{\gamma(a) : \gamma \in \Gamma\}$. The following result is fundamental to the theory. The proof is straightforward and may be found in [Ahlfors, 1981, p.79].

Theorem 1.4.1   If Γ is a discrete group preserving B and $a \in B$ then $\Gamma(a)$ can accumulate only at S.

If we select $a, b \in B$ and a sequence of Moebius transforms $\{\gamma_n\}$ with $\gamma_n(a) \rightarrow \xi \in S$ then, since Moebius transforms preserve hyperbolic distance, it is easy to see that $\gamma_n(b) \rightarrow \xi$ also. In view of this remark the following definition makes sense.

A point $\xi \in S$ is a **limit point** for the discrete group Γ if for one, and hence every, point $x \in B$ the orbit $\Gamma(x)$ accumulates at ξ. The set of limit points is denoted by $\Lambda(\Gamma)$, or simply Λ. The next chapter is devoted to a detailed discussion of Λ and some interesting subsets. For the time being we merely remark that Λ is a closed subset of S and its complement is the set of **ordinary points**. The group Γ is said to be of the **first kind** if $\Lambda = S$ and of the **second kind** otherwise.

A **Fuchsian group** is a discrete subgroup of $M(B)$ in dimension $n = 2$ and a **Kleinian group** is a discrete subgroup of $M(B)$ in dimension $n = 3$. This is the modern terminology — in much classical work the word "Kleinian" was used for a discrete group of Moebius transforms acting on the complex plane with a non-empty ordinary set. Poincaré showed how to extend the action of such a group to the upper half space in R^3 (with the complex plane viewed as the plane $x_3 = 0$) and it becomes (in our terminology) a Kleinian group of the second kind — see [Poincaré, 1883].

If Γ is a discrete subgroup of $M(B)$ then two points $a, \gamma(a)$, $\gamma \in \Gamma$ are called equivalent, and we can pass to the quotient B/Γ by identification of equivalent points. This quotient is certainly a Hausdorff space, and in dimensions 2 and 3 it is known to be a manifold. In higher dimensions one encounters real difficulty with the fixed points, and the more general notions of V-manifold (Satake) and orbifold (Thurston) must be used. The reader is referred to [Satake, 1956] and [Davis and Morgan, 1984 p.183] for full details. For our purposes, we shall simply refer to B/Γ as the quotient space.

A convenient way to view the quotient space B/Γ is via the notion of a **fundamental region**. The basic idea is to select a representative from each orbit and to do this in such a way that the resulting set has nice geometric properties. A detailed account of such constructions, for Fuchsian groups, is to be found in [Beardon, 1983, Chapter 9]. For our purposes we restrict attention to the most well known such construction. If $a \in B$ and Γ is a discrete group preserving B then set

$$D_a = \text{interior } \{x \in B : \rho(x,a) \leq \rho(x,\gamma(a)) \ \gamma \in \Gamma\}.$$

The region D_a is convex in the hyperbolic sense (being the intersection of half-spaces), it is open, and any point x of B is equivalent to some point of its closure. The stabilizer of a is a finite subgroup of Γ, and if D_a is intersected with a fundamental domain for this stabilizer, then a fundamental domain for Γ results. In particular, if a is fixed only by the identity element of Γ then D_a is a fundamental domain for Γ — from each orbit we have selected the point closest to a. In this case D_a is called the *Dirichlet region* centered at a. The boundary of D_a is a countable collection of $(n-1)$-dimensional faces which occur in Γ-equivalent pairs. Further, the face pairing transforms are a set of generators for the group.

We shall be concerned with the notion of **geometrically finite groups**. In dimensions 2 and 3 the standard definition of geometrical finiteness is that some Dirichlet region D_a have finitely many faces. In [Marden, 1974] it is shown that in this case every Dirichlet region, indeed every convex fundamental polyhedron, has finitely many faces. Geometrical finiteness in dimensions 2 and 3 clearly implies that the group is finitely generated. The converse is true in dimension 2, [Beardon, 1983, p.254], and is false in dimension 3 [Greenberg, 1966]. In higher dimensions there are problems with the definition — examples are known of discrete groups in dimension 4 for which some convex polyhedron has finitely many faces and some other does not. The reader is referred to [Apanasov, 1982], [Apanasov, 1983], and to [Bowditch, 1988] for an account of such phenomena. For the present we simply adopt the definition that a group Γ is **geometrically finite** if **some** convex fundamental polyhedron (not necessarily a Dirichlet domain) has finitely many faces. We will return to this topic at the end of Chapter 2.

If Λ is the limit set of the group Γ we may form the convex hull of Λ — $C(\Lambda)$ in B. The set $C(\Lambda)$ is invariant under Γ and we say that Γ is **convex co—compact** if the action of Γ on the convex hull $C(\Lambda)$ has a compact fundamental region in B. As an example, geometrically finite Fuchsian groups without parabolic elements are convex co-compact.

From the differential $d\rho$ defined in B we may construct a hyperbolic volume element as follows

$$dV = \frac{2^n \, dx_1 dx_2 \dots dx_n}{(1 - |x|^2)^n}$$

and use this to measure the volume of a Dirichlet region D_a. The discrete group Γ is said to be of **finite volume** if for one (and hence every) non-fixed point

$a \in B$, $V(D_a) < \infty$. As examples, one can consider any group with D_a compact in B or a geometrically finite Fuchsian group of the first kind.

1.5 The Orbital Counting Function

We are interested in the distribution of orbits in the ball and define an orbital counting function as follows. If $x, y \in B$ and r is positive set

$$N(r, x, y) \; = \; \text{card } \{\gamma \in \Gamma : \rho(x, \gamma(y)) < r\} \tag{1.5.1}$$

The next result generalizes an old estimate, usually attributed to Tsuji, but in fact first established by Hopf [Hopf, 1936] which states that for a Fuchsian group there is a constant A such that $N(r, x, y) < Ae^r$.

Theorem 1.5.1   Let Γ be a discrete group acting in B. There is a constant A depending on Γ, and y such that for any $x \in B$

$$N(r, x, y) \; < \; Ae^{r(n-1)}.$$

Proof. We start by calculating the hyperbolic volume of a ball of hyperbolic radius s. We may suppose that the ball is centered at the origin. If $|x| = t$ then $\rho(0, x) = \log \left[\dfrac{1+t}{1-t} \right]$ and we have

$$|x| = \tanh \left[\rho(0, x)/2 \right]. \tag{1.5.2}$$

Using the Euclidean volume element computed in section 1.1 we see that

$$V\{x : \rho(x, 0) < s\} \; = \; W \int_0^{\tanh s/2} \frac{2^n |x|^{n-1}}{(1 - |x|^2)^n} \, d|x|$$

where W is the w-measure of the unit sphere S. From (1.5.2) and the standard identities for hyperbolic functions we obtain

$$V\{x : \rho(x, 0) < s\} \; = \; W \int_0^s \sinh^{n-1}(t) dt. \tag{1.5.3}$$

Let Δ be a ball centered at y of hyperbolic radius $\epsilon > 0$ so small that no two Γ-images of Δ overlap — note that ϵ is a function of the minimal separation of the orbit points of y. Since the images of Δ do not overlap we have

$$V(\Delta)N(r, x, y) < V\{z : \rho(x, z) < r + \epsilon\} = W \int_0^{r+\epsilon} \sinh^{n-1}(t) dt$$

from (1.5.3). The integral is clearly bounded above by

$$\frac{1}{2^{n-1}} \int_0^{r+\epsilon} e^{(n-1)t} \, dt$$

from which we obtain

$$N(r,x,y) \; < \; \frac{e^{\epsilon(n-1)}}{2^n(n-1)V(\Delta)} \, e^{(n-1)r}$$

and this is the required result. $\square$

In the case that Γ is of finite volume we have an inequality in the opposite direction.

Theorem 1.5.2   Let Γ be a discrete group acting in B. If Γ is of finite volume then for $x,y \in B$ there is a constant C depending on x,y, and on Γ and a positive real r_0 such that if $r > r_0$

$$N(r,x,y) \; > \; C \, e^{(n-1)r}.$$

Proof. We prove the result with $x = 0$. Consider the ball $\Delta(0,r)$ centered at the origin and of hyperbolic radius r. Set $D(r) = D \cap \Delta(0,r)$ where D is the Dirichlet region centered at the origin. If the origin is fixed by some (necessarily elliptic) element then the Dirichlet construction yields a cover of finite volume and the argument which follows goes through with minor modifications. Since $V(D) < \infty$, for $\epsilon > 0$ we may find r_0 such that for $r \geq r_0$

$$V(D - D(r_0)) \; < \; \epsilon. \tag{1.5.4}$$

Now for $r > r_0$

$$V(\Delta(0,r)) \; = \; V[\Delta(0,r) \cap \Gamma(D(r_0))] + V[\Delta(0,r) \cap \Gamma(D - D(r_0))] \tag{1.5.5}$$

and the second term on the right is equal to

$$\int_{D - D(r_0)} N(r,0,y)\,dV(y).$$

However, from theorem 1.5.1 and inequality (1.5.4), this integral does not exceed $A\,\epsilon e^{(n-1)r}$. Using this estimate, and the formula (1.5.3) for the volume of a hyperbolic ball, we see from (1.5.5) that

$$V[\Delta(0,r) \cap \Gamma(D(r_0))] \; > \; \int_0^r \sinh^{n-1}(t)\,dt \; - \; A\,\epsilon e^{(n-1)r} \; > \; A_1 e^{(n-1)r}$$

for some positive A_1 — provided ϵ was chosen small enough. We assume, without loss of generality, that y belongs to $D(r_0)$. The Γ-images of $D(r_0)$ are disjoint and if one of them meets $\Delta(0,r)$ then the corresponding image of y must lie in $\Delta(0,r+2r_0)$ and so we have

$$V(D(r_0))\, N(r+2r_0,0,y) \; > \; A_1 e^{(n-1)r}.$$

From which we obtain

$$N(r,0,y) \; > \; \frac{A_1}{V(D)\, e^{2r d(n-1)}} \; e^{(n-1)r}$$

as required. □

The orbital counting function is a notion of central importance in the theory of discrete groups. It has been studied extensively and deep results have been obtained. We will be considering this function in greater detail in later chapters and will derive, for example, some asymptotic results relating it to the so called "critical exponent" of the group and to the Hausdorff dimension of the limit set. The two results given in this section are the minimum necessary to study the convergence questions of the next section and the following chapter.

1.6 Convergence Questions

We are interested in the rate at which orbit tends to S. The first observation is that any two orbits $\Gamma(a)$, $\Gamma(b)$ are comparable in the sense that the ratios

$$\frac{1 - |\gamma(a)|}{1 - |\gamma(b)|} \; : \; \gamma \in \Gamma$$

lie between finite limits. To see this, note that $\rho(0,\gamma b) \leq \rho(0,\gamma a) + \rho(a,b)$ and so

$$\log \left[\frac{1 + |\gamma(b)|}{1 - |\gamma(b)|} \right] \leq \log \left[\frac{1 + |\gamma(a)|}{1 - |\gamma(a)|} \right] + \rho(a,b)$$

from which

$$1 - |\gamma(a)| \leq 2 e^{\rho(a,b)} (1 - |\gamma(b)|)$$

for all γ.

A good way to study the rate at which orbits tend to S is to consider the convergence of the series

$$\sum_{\gamma \in \Gamma} (1 - |\gamma(a)|)^{\alpha}$$

for various $\alpha > 0$. From our remarks above, the convergence or otherwise of this series is independent of $a \in B$. Thus in general we will consider the series

$$\sum_{\gamma \in \Gamma} (1 - |\gamma(0)|)^{\alpha}. \tag{1.6.1}$$

In many ways it is more natural to look instead at the series

$$\sum_{\gamma \in \Gamma} e^{-\alpha \rho(0,\gamma(0))}. \tag{1.6.2}$$

and in view of the fact that $\rho(0,\gamma(0)) = \log \left[\frac{1 + |\gamma(0)|}{1 - |\gamma(0)|} \right]$ it is immediate that

(1.6.1) and (1.6.2) converge or diverge together.

For $\alpha > n-1$, the series always converges.

Theorem 1.6.1   Let Γ be a discrete group preserving B and $\alpha > n-1$ then

$$\sum_{\gamma \in \Gamma} e^{-\alpha\rho(0,\gamma(0))} < \infty.$$

Proof. We consider the partial sum

$$\sum_{\substack{\gamma \in \Gamma \\ \rho(0,\gamma(0)) < R}} e^{-\alpha\rho(0,\gamma(0))} = \int_0^R e^{-\alpha t}\, dN(t,0,0) = N(R,0,0)\, e^{-\alpha R} + \alpha\int_0^R N(t,0,0)e^{-\alpha t}\, dt$$

and the result follows immediately from theorem 1.5.1 □

We next define the **critical exponent** of a discrete group. Set

$$\delta(\Gamma) = \inf\{\alpha : \sum_{\gamma \in \Gamma} e^{-\alpha\rho(0,\gamma(0))} < \infty\}.$$

This is a construct of fundamental importance in our later work (see Chapter 3). We note from theorem 1.6.1 that, for any discrete group Γ, $\delta \leq n-1$. The group Γ is said to be of **convergence** or **divergence** type according as the series

$$\sum_{\gamma \in \Gamma} e^{-\delta(\Gamma)\rho(0,\gamma(0))} \qquad (1.6.3)$$

converges or diverges. We remark that this definition is different from the usual one. It is usual to say that Γ is of convergence type if the series

$$\sum_{\gamma \in \Gamma} e^{-(n-1)\rho(0,\gamma(0))} \qquad (1.6.4)$$

converges and of divergence type otherwise.

The convergence of the series (1.6.4) is, as we shall see, necessary and sufficient for the existence of a Green's function on the quotient space B/Γ and it is for this reason that the convergence or otherwise of the series (1.6.4) is viewed as an important dichotomy. For our purposes, the convergence or otherwise of the series (1.6.3) is an even more important dichotomy and pervades the entire measure construction given in Chapter 3. It is for this reason that we adopt the non-standard notation. To reiterate, convergence or divergence type will refer to the series (1.6.3). If the series (1.6.4) converges we will say "Γ converges at the exponent $n-1$".

Theorem 1.6.2   Let Γ be a discrete group preserving B. If Γ is of the second kind then Γ converges at the exponent $n-1$.

Proof. The limit set is a closed subset of S and so there is a subset C of ordinary points of S which is the intersection of S and a ball in R^n. Since the closure of C is compact, it follows that C can only meet finitely many of its Γ-images and consequently

$$\sum_{\gamma \in \Gamma} w(\gamma(C)) < \infty$$

which implies

$$\sum_{\gamma \in \Gamma} \int_C |\gamma'(x)|^{n-1} dw(x) < \infty. \tag{1.6.5}$$

Using theorem 1.3.4 and the fact that $|\gamma(0)| = |\gamma^{-1}(0)|$ we see that (1.6.5) implies

$$\sum_{\gamma \in \Gamma} \int_C (1 - |\gamma(0)|^2)^{n-1} |x - \gamma^{-1}(0)|^{-2(n-1)} dw(x) < \infty \tag{1.6.6}$$

but, since $|x - \gamma^{-1}(0)| < 2$ for all $\gamma \in \Gamma$ and $x \in C$, the convergence of the series

$$\sum_{\gamma \in \Gamma} (1 - |\gamma(0)|)^{n-1}$$

follows from (1.6.6) above. $\square$

Theorem 1.6.3   Let Γ be a discrete group preserving B. If Γ is of finite volume then $\delta(\Gamma) = n-1$ and Γ is of divergence type.

Proof. As in the proof of theorem 1.6.1 consider the partial sum

$$\sum_{\substack{\gamma \in \Gamma \\ \rho(0,\gamma(0)) < R}} e^{-(n-1)\rho(0,\gamma(0))} = \int_0^R e^{-(n-1)t} \, dN(t,0,0)$$

$$= N(R,0,0) \, e^{-(n-1)R} + (n-1) \int_0^R N(t,0,0) e^{-(n-1)t} \, dt$$

which, from theorem 1.5.2 , clearly diverges as $R \to \infty$. This, with theorem 1.6.1, completes the proof. $\square$

CHAPTER 2

The Limit Set

2.1 Introduction

Suppose that Γ is a discrete group of Moebius transformations preserving B. Given $x \in B$ the orbit of x under Γ, written $\Gamma(x)$, is defined by

$$\Gamma(x) = \{\gamma(x) : \gamma \in \Gamma\}.$$

Discreteness implies that such an orbit can accumulate only at S (theorem 1.4.1). The subset $\Lambda(\Gamma)$ of S at which orbits accumulate is the limit set of Γ. Our aim in this chapter is to study various classes of limit points which arise in a natural way in the study of ergodic properties. This classification is achieved by studying the rate at which orbits approach the limit point. In order to unify our treatment and, at a later stage, to make various analogies and connections clearer it will help at this point to make a definition.

For a discrete group Γ enumerated by $\Gamma = \{\gamma_n : n = 0,1,2,...\}$ we define the following subset of the unit sphere S

$$L(a:k,\alpha) = \bigcap_{N=1}^{\infty} \bigcup_{n > N} I(\gamma_n(a):k,\alpha). \qquad (2.1.1)$$

Where $I(a:k,\alpha)$ is the set defined by (1.2.4). Thus $L(a:k,\alpha)$ comprises those points of S which lie in infinitely many of the neighborhoods $I(\gamma(a):k,\alpha)$, $\gamma \in \Gamma$. Since, on any sequence $\{\gamma_m\}$, $|\gamma_m(a)| \to 1$ we see that, for any $k,\alpha > 0$, $L(a:k,\alpha)$ comprises limit points. Note further that the size of k and α regulate the rate at which the orbit of a approaches $\xi \in L(a:k,\alpha)$.

Our next result is reminiscent of the Borel Cantelli lemma from probability theory.

Theorem 2.1.1   Let Γ be a discrete group acting in B for which the series

$$\sum_{\gamma \in \Gamma} (1 - |\gamma(a)|)^{(n-1)\alpha}$$

converges. Then the set $\bigcup_{k > 0} L(a:k,\alpha)$ has zero w - measure as a subset of S.

Proof. Fix $k > 0$. From lemma 1.1.1 we observe that

$$w(I(\gamma(a):k,\alpha)) \le \frac{M}{n-1}\mu^{n-1}$$

where $\mu = \arccos (1 - k^2(1 - |\gamma(a)|)^{2\alpha}/2)$. Thus for $|\gamma(a)|$ close enough to 1 we will have $\mu < 2k(1 - |\gamma(a)|)^{\alpha}$, and so, except for finitely many $\gamma \in \Gamma$, we will have

$$w(I(\gamma(a):k,\alpha)) \le A \ (1 - |\gamma(a)|)^{(n-1)\alpha} \qquad (2.1.2)$$

where A is a constant depending only upon k and the dimension n. If $w(L(a:k,\alpha)) = \epsilon > 0$ then infinitely many of the sets $I(\gamma(a):k,\alpha)$ would have w-measure at least ϵ and we deduce from (2.1.2) that

$$\sum_{\gamma \in \Gamma} (1 - |\gamma(a)|)^{(n-1)\alpha} = \infty.$$

This contradiction shows that $w(L(a:k,\alpha)) = 0$ for any $k > 0$. Now from the definition (1.2.4) we see that if $k' > k > 0$ then $I(a:k,\alpha) \subset I(a:k',\alpha)$ and so the set $\bigcup_{\gamma \in \Gamma} L(a:k,\alpha)$ may be written as a countable union of sets of zero w-measure and this completes the proof of the theorem. $\square$

Following ideas of Sprindzuk [Sprindzuk, 1979, p.21] we will prove

Theorem 2.1.2   Let Γ be a discrete group acting in B. Fix $\alpha > 0$ and k_1, k_2 satisfying $k_1 > k_2 > 0$ then, for any $a \in B$,

$$w(L(a:k_1,\alpha)) = w(L(a:k_2,\alpha)).$$

Proof. Note that for any $\gamma \in \Gamma$, $I(\gamma(a):k_2,\alpha) \subset I(\gamma(a):k_1,\alpha)$. If we write $\Gamma = \{\gamma_m : m = 0,1,2,...\}$ then, as $m \to \infty$,

$$w(I(\gamma_m(a):k_1,\alpha)) \sim \frac{M \ k_1^{n-1}}{n-1}(1 - |\gamma_m(a)|)^{(n-1)\alpha}$$

from lemma 1.1.1. It follows that there exists $\delta > 0$ such that, for m large enough,

$$\frac{w(I(\gamma_m(a):k_2,\alpha))}{w(I(\gamma_m(a):k_1,\alpha))} > \delta. \tag{2.1.3}$$

Now define

$$J = \bigcap_{l-1}^{\infty} \bigcup_{m-l}^{\infty} I(\gamma_m(a):k_1,\alpha) \quad \text{and} \quad B_l = \bigcup_{m-l}^{\infty} I(\gamma_m(a):k_2,\alpha)$$

and set $D_l = J - B_l$. To prove the theorem it suffices to show that every D_l is of w-measure zero. If this is not the case then D_l contains a point of metric density — say ξ. Since $\xi \in J$ then $\xi \in I(\gamma_m(a):k_{1,}\alpha)$ for infinitely many m and, for such m,

$$w[D_m \cap I(\gamma_m(a):k_1,\alpha)] \sim w(I(\gamma_m(a):k_1,\alpha)) \quad \text{as} \quad m \to \infty \tag{2.1.4}$$

since $w(I(\gamma_m(a):k_1,\alpha)) \to 0$ as $m \to \infty$. On the other hand, the sets D_m and $I(\gamma_m(a):k_2,\alpha)$ do not intersect if $m \geq l$, and hence $D_m \cap I(\gamma_m(a):k_1,\alpha)$ and $I(\gamma_m(a):k_2,\alpha)$ are non-intersecting subsets of $I(\gamma_m(a):k_1,\alpha)$. Therefore

$$w[I(\gamma_m(a):k_1,\alpha)] \geq w[I(\gamma_m(a):k_2,\alpha)] + w[D_m \cap I(\gamma_m(a):k_1,\alpha)]$$

$$\geq \delta\, w[I(\gamma_m(a):k_1,\alpha)] + w[D_m \cap I(\gamma_m(a):k_1,\alpha)]$$

— by (2.1.3). It follows that $w[D_m \cap I(\gamma_m(a):k_1,\alpha)] \leq (1 - \delta)\, w[I(\gamma_m(a):k_1,\alpha)]$ which contradicts (2.1.4). This completes the proof of the theorem. $\square$

The following corollary is immediate.

Corollary 2.1.3   Let Γ be a discrete group acting in B. Fix $\alpha > 0$ and $a \in B$ then

$$w(\bigcup_{k>0} L(a:k,\alpha)) = w(\bigcap_{k>0} L(a:k,\alpha)).$$

Our analysis of the limit set will be based upon the rate at which orbits approach the point in question. We will start by considering the most rapid rate possible and then successively weaken the required rate of approach.

2.2 The Line Transitive Set

Given a discrete group Γ acting in B and a point $\xi \in \Lambda(\Gamma)$ then, for any $\gamma \in \Gamma$ and any $a \in B$ we clearly have $1 - |\gamma(a)| \leq |\xi - \gamma(a)|$. In terms of orbital approach, the best we can hope for is that, on a sequence $\{\gamma_n\} \subset \Gamma$,

$$\frac{|\xi - \gamma_n(a)|}{1 - |\gamma_n(a)|} \to 1 \qquad \text{as} \quad n \to \infty.$$

We could even ask that for any $a \in B$ such a sequence $\{\gamma_n\}$ exist. In fact we start by asking even more than this.

Definition. The point $\xi \in \Lambda(\Gamma)$ is said to be a line transitive point for Γ if for every pair $a, b \in B$ there exists a sequence $\{\gamma_n\} \subset \Gamma$ such that

$$\lim_{n \to \infty} \frac{|\xi - \gamma_n(a)|}{1 - |\gamma_n(a)|} = 1 \quad \text{and} \quad \lim_{n \to \infty} \frac{|\xi - \gamma_n(b)|}{1 - |\gamma_n(b)|} = 1.$$

Suppose ξ is line transitive and σ is a geodesic ending at ξ (with η the other end point of σ) we have

$$\cosh \rho(\gamma_n(a), \sigma) = \frac{2|\gamma_n(a) - \xi||\gamma_n(a) - \eta|}{|\xi - \eta|(1 - |\gamma_n(a)|^2)}$$

(from theorem 1.2.1) and so, on the sequence $\{\gamma_n\}$, $\rho(\gamma_n(a), \sigma) \to 0$ and, similarly, $\rho(\gamma_n(b), \sigma) \to 0$. By the invariance of the hyperbolic metric we have

$$\rho(a, \gamma_n^{-1}(\sigma)) \to 0 \quad \text{and} \quad \rho(b, \gamma_n^{-1}(\sigma)) \to 0$$

as $n \to \infty$. Thus, for any pair of points $a, b \in B$ there is a sequence of images of the geodesic σ coming arbitrarily close to both points. We have proved the following result.

Theorem 2.2.1   If the point $\xi \in \Lambda(\Gamma)$ is a line transitive point and σ is an arbitrary geodesic ending at ξ then the Γ-images of σ are dense in the set of all geodesics.

This result explains the name "line transitive" - the set of line transitive points is denoted by T_l. The class T_l was the first special class of limit points to be isolated. Artin [Artin, 1924] characterized T_l for the modular group acting in the upper half of the complex plane — he showed that T_l comprises those real numbers whose continued fraction representation contains each finite sequence of integers. Myrberg [Myrberg, 1931] later showed that for finitely generated Fuchsian groups of the first kind, the set T_l has full measure on the circle. Other early work on the set T_l (in dimension 2) is to be found in: [Koebe, 1930], [Lobell, 1929], [Myrberg, 1931], and [Shimada, 1960]. The papers of Koebe and Lobell contain a proof of the following result. Since the original papers are hard to find, a proof is included for the sake of completeness.

Theorem 2.2.2   If Γ is of the first kind then $T_l \neq \emptyset$.

Proof. Since Γ is of the first kind then $\Lambda(\Gamma) = S$ and we know that the set of hyperbolic fixed point pairs is dense in $S \times S$ [Gottschalk and Hedlund, 1955, p.122]. Following Hedlund's methods [Gottschalk and Hedlund, 1955, p.123] it may be shown that if A,B,C,D are any four open neighborhoods in S then there exists $\gamma \in \Gamma$ with $\gamma(A)\cap C \neq \varnothing$ and $\gamma(B)\cap D \neq \varnothing$.

Now for integer n we may partition S into n regions of equal w - measure, say $E_1, \ldots, E_n$. Choose A,B open neighborhoods in S and select E_i, E_j. By the continuity of Moebius transforms we may find open sub-neighborhoods, say A',B', of A,B and a Moebius $\gamma_{i,j} \in \Gamma$ such that if $a \in A'$, $b \in B'$ then $\gamma_{i,j}(a) \in E_i$, $\gamma_{i,j}(b) \in E_j$. This procedure may be repeated for all pairs E_i, E_j and we have two open neighborhoods — say A,B of S and a collection $\gamma_{i,j}$ of Moebius transforms in Γ such that for any $a \in A$, $b \in B$, $\gamma_{i,j}(a) \in E_i$, $\gamma_{i,j}(b) \in E_j$.

We repeat this procedure with the integer $n+1$, starting with the neighborhoods A,B just obtained, and find ultimately that there exists a geodesic whose Γ-images are dense in the set of all geodesics. One end point of this geodesic must be in T_l and the theorem is proved. □

2.3 The Point Transitive Set

If we weaken the requirement for a line transitive point and require only that for every $a \in B$ a sequence of Γ-images of a approach the limit point almost radially, then the limit point is said to be point transitive.

Definition. The point $\xi \in \Lambda(\Gamma)$ is said to be a point transitive point for Γ if for every $a \in B$ there exists a sequence $\{\gamma_n\} \subset \Gamma$ such that

$$\lim_{n \to \infty} \frac{|\xi - \gamma_n(a)|}{1 - |\gamma_n(a)|} = 1.$$

The argument used in section 2.2 easily yields the following result which explains the name "point transitive".

Theorem 2.3.1   If the point $\xi \in \Lambda(\Gamma)$ is a point transitive point and σ is an arbitrary geodesic ending at ξ then the Γ-images of σ are dense in B.

The set of point transitive points will be denoted T_p and clearly $T_l \subset T_p$ for any discrete group Γ. In [Sheingorn, 1980a] it is shown that in general these sets are not equal.

Theorem 2.3.2   (Sheingorn) If Γ is the modular group acting in the upper half of the complex plane then $T_p \neq T_l$.

Whereas, for groups of the first kind, the set T_l is always nonempty, for groups of the second kind, the set T_p is always empty.

Theorem 2.3.3   If Γ is of the second kind then $T_p = \emptyset$.

Proof. Let $\eta \in S$ be an ordinary point for Γ. Since the ordinary set is open, there exists a neighborhood N of η in S which comprises ordinary points. It is geometrically evident that a point a of B may be chosen so close to η that any geodesic passing through a has one end point in N. Such a geodesic clearly cannot be approximated by images of any other geodesic since this would make the end point in N a limit point for the group. $\square$

The class T_p has been used extensively in connection with number theory and the boundary behavior of automorphic functions and forms — see, for example: [Lehner, 1964 Chapter 10], [Nicholls, 1981], and [Sheingorn, 1980b].

2.4 The Conical Limit Set

We next weaken the requirement that orbits approach a limit point almost radially and require instead that they approach within a cone. In view of theorem 1.2.4 we see that the property given in the following definition is equivalent to conical approach.

Definition. The point $\xi \in \Lambda(\Gamma)$ is said to be a conical limit point for Γ if for every $a \in B$ there exists a sequence $\{\gamma_n\} \subset \Gamma$ on which the sequence $\dfrac{|\xi - \gamma_n(a)|}{1 - |\gamma_n(a)|}$ remains bounded.

It is immediate from the definition that the conical limit set — denoted by C — is a subset of T_p. The following result is an immediate consequence of theorem 1.2.4.

Theorem 2.4.1   The point $\xi \in S$ is a conical limit point for Γ if and only if there is a geodesic σ ending at ξ such that for any point $a \in B$ there are infinitely many Γ-images of σ within a bounded hyperbolic distance of a.

Corollary 2.4.2   If ξ is fixed by a loxodromic element of Γ then ξ is a conical limit point.

It is well known that the closure of the set of group images of the axis of a loxodromic element in the group comprises the set of images of the axis and so, from theorem 2.3.1, a loxodromic fixed point is not in T_p. Thus $T_p - C$ is not empty. However, this difference is fairly small.

Theorem 2.4.3   If Γ is a discrete group then $w(C) = w(T_p)$.

Proof. Recalling the set $L(x{:}k,\alpha)$ defined in (2.1.1) and using theorem 1.2.4 we see that if $\{x_n\}$ is a countable dense subset of B we have

$$T_p = \bigcap_{n\geq 1}\bigcap_{k>0} L(x_n:k,1) \quad \text{and} \quad C = \bigcap_{n\geq 1}\bigcup_{k>0} L(x_n:k,1)$$

and so $w(T_p) = w(C)$ from corollary 2.1.3. □

Using the fact that

$$C = \bigcap_{x\in B}\bigcup_{k>0} L(x:k,1) = \bigcup_{k>0} L(x:k,1)$$

for any $x \in B$, the following result is a corollary of theorem 2.1.1.

Theorem 2.4.4   Let Γ be a discrete group acting in B for which the series

$$\sum_{\gamma\in\Gamma}(1 - |\gamma(a)|)^{n-1}$$

converges. Then $w(C) = 0$.

Conical limit points were introduced (in dimension 2) by Hedlund [Hedlund, 1936] and were used by him in his study of horocyclic transitive points. The conical limit set has been studied over the years by a number of authors. Particular mention should be made of : [Lehner, 1964, Chapter 10] where the connection with Diophantine approximation is made; [Beardon and Maskit, 1974] for the characterization (2) of theorem 1.2.4 and the generalization of Hedlund's results to the three dimensional case; and [Agard, 1983] and [Tukia, 1985] for applications in the development of general rigidity theorems.

We next characterize the conical limit set in terms of shadows — see section 1.2 for the definition. Suppose $\{a_n\}$ is a sequence of points in B such that $|a_n| \to 1$ as $n \to \infty$ and $\delta > 0$ is chosen. From (1.2.1) it follows that $\xi \in S$ belongs to the shadows $b(0:a_n,\delta)$, $n = 1,2,...$ if and only if there is a constant $k > 0$ such that for n large enough $|\xi - a_n| < k\,(1 - |a_n|)$. This implies, by theorem 1.2.4, that the sequence $\{a_n\}$ converges to ξ in a cone. We have proved the following.

Theorem 2.4.5   Let Γ be a discrete group acting in B and $\xi \in S$. Then ξ is a conical limit point for Γ if and only if for some $a \in B$ and $\delta > 0$ ξ belongs to infinitely many shadows $b(0:\gamma(a),\delta) : \gamma \in \Gamma$.

Of critical importance for the ergodic theory is the following theorem. It follows from a deep ergodic result [Sullivan, 1981 p.483], however, we prefer to give an elementary proof. This proof is taken directly from [Ahlfors, 1981] and is attributed by him to Thurston.

Theorem 2.4.6   Let Γ be a discrete group acting in B for which the set of conical limit points has zero w-measure. Then the series

$$\sum_{\gamma \in \Gamma} (1 - |\gamma(a)|)^{n-1}$$

converges for all $a \in B$.

Proof. If the group Γ is enumerated by $\Gamma = \{\gamma_p : p = 0,1,2,...\}$ we write $x_p = \gamma_p^{-1}(0)$ and note from theorem 1.2.4 that

$$C = \bigcup_{\delta > 0} \bigcap_{N=1}^{\infty} \bigcup_{p \geq N} b(0:x_p,\delta) .$$

From this it is immediate that if $w(C) = 0$ then, for every $\delta > 0$,

$$\lim_{N \to \infty} w\big(\bigcup_{p \geq N} b(0:x_p,\delta)\big) = 0. \qquad (2.4.1)$$

For the group Γ to converge at exponent $n-1$ it is necessary and sufficient, by theorem 1.2.2, that

$$\sum_{p=1}^{\infty} w\big(b(0:x_p,\delta)\big) < \infty. \qquad (2.4.2)$$

It is evident that (2.4.1) implies (2.4.2) provided the shadows $b(0:x_p,\delta)$ do not overlap too much. The idea of the proof is to show that this is the case.

As a first simplification we show that many of the shadows can be discarded. For this purpose choose a number $A > 0$, which will ultimately be large, and use it to define a subsequence $\{p_k\}$ as follows:

• choose $p_0 = 0$

• suppose $p_0,...,p_k$ have been chosen so that the distances $\rho(x_{p_i},x_{p_j})$ are all greater than A. Then choose p_{k+1} to be the smallest p such that $\rho(x_{p_i},x_{p_{k+1}}) > A$ for $i = 0,...,k$.

This choice can always be made and we see that

$$\rho(x_{p_h},x_{p_k}) > A \qquad \text{for all } h \neq k$$

and for every p there exists a p_k such that

$$\rho(x_p,x_{p_k}) \leq A.$$

Let N be the number of x_p with $\rho(x_p,0) \leq A$ and note that this is also the number of x_p with $\rho(x_p,x_{p_k}) \leq A$. If this is the case then

$$\frac{1 - |x_p|}{1 - |x_{p_i}|} \le P$$

where P depends only on A. If

$$\sum_{k=1}^{\infty} (1 - |x_{p_k}|)^{n-1} < \infty$$

it follows that

$$\sum_{k=1}^{\infty} (1 - |x_p|)^{n-1} < \infty.$$

Thus it is sufficient to prove (2.4.2) on the subsequence, and we renumber the subsequence as $\{x_p : p = 0,1,2,...\}$.

Choose A so large that the balls $\Delta(x_p,\delta)$ do not overlap. Imagine an observer placed at the origin and we speak of total or partial eclipses when two shadows overlap. We next divide the $\Delta(x_p,\delta)$ into classes depending on the number of times they are eclipsed.

The class I_0 consists only of $\Delta(0,\delta)$. We remove this ball and define I_1 as the class of all $\Delta(x_p,\delta)$ that are now completely visible from the origin. Next remove all balls of I_1 and define I_2 as the class of those $\Delta(x_p,\delta)$ which are now completely visible. By induction, the class I_n is defined for $n = 0,1,2,....$ Clearly each ball $\Delta(x_p,\delta)$ belongs to a class I_n and the shadows of the balls in I_n are disjoint.

The object of the proof is to show that for some λ, $0 < \lambda < 1$,

$$\sum_{\Delta(x_p,\delta) \in I_{m+1}} w(b(0:x_p,\delta)) \le \lambda \sum_{\Delta(x_p,\delta) \in I_m} w(b(0:x_p,\delta))$$

which clearly implies (2.4.2).

Every $\Delta(x_p,\delta) \in I_{m+1}$ is either partially or totally eclipsed by some $\Delta(x_q,\delta) \in I_m$ and writing r_p, r_q for the Euclidean radii of $\Delta(x_p,\delta)$, $\Delta(x_q,\delta)$ we need an upper bound for the ratio r_p / r_q. We refer to figure 2.4.1 in which x_p and x_q are at a non-Euclidean distance $< \delta$ from the same radius. Let b_p, b_q be the non-Euclidean orthogonal projections of x_p, x_q onto this radius. Then

$$A < \rho(x_p,x_q) < \rho(b_p,b_q) + 2\delta$$

$$= \rho(0,b_p) - \rho(0,b_q) + 2\delta$$

$$< \rho(0,x_p) - \rho(0,x_q) + 4\delta$$

or

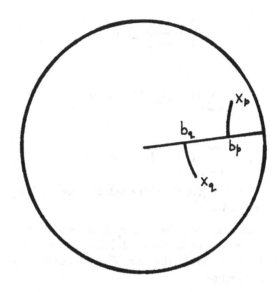

Figure 2.4.1

$$\rho(0,x_p) \;>\; \rho(0,x_q) + A - 4\delta$$

and

$$\frac{1+|x_p|}{1-|x_p|} \;>\; \frac{1+|x_q|}{1-|x_q|}\, e^{A-4\delta}$$

which implies

$$1 - |x_p| \;<\; 2\,e^{-A+4\delta}(1 - |x_q|) \,.$$

The formula for the Euclidean radius r of the ball $\Delta(x,\delta)$ is

$$r \;=\; \frac{(1-|x|^2)\tanh \delta/2}{1-|x|^2 \tanh^2 \delta/2}$$

(see [Ahlfors, 1981 p.86]) and thus we find

$$r_p / r_q \;<\; 4\,e^{-A+4\delta}\cosh^2(\delta/2) \tag{2.4.3}$$

which can be made arbitrarily small by choosing A large enough.

We consider now all the $\Delta(x_p,\delta) \in I_{m+1}$ which are partially eclipsed by $\Delta(x_q,\delta) \in I_m$. The shadows $b(0\!:\!x_p,\delta)$ are disjoint and lie asymptotically within a distance r_p from the rim of $b(0\!:\!x_q,\delta)$. Their total area is therefore asymptotically at most the difference in areas of two spherical caps of radii $r_p + r_q$ and $r_q - r_p$. This quantity is seen from lemma 1.1.1 to be

$$M \int_0^{\mu_1} (\sin \theta)^{n-2} \, d\theta - M \int_0^{\mu_2} (\sin \theta)^{n-2} \, d\theta$$

where $\mu_1 = \arccos [1 - (r_p + r_q)^2 / 2], \quad \mu_2 = \arccos [1 - (r_q - r_p)^2 / 2] \quad$ and asymptotically this is

$$\frac{M (r_p + r_q)^{n-1}}{n-1} - \frac{M (r_q - r_p)^{n-1}}{n-1} = \frac{M}{n-1} r_q^{n-1} \left[(1 + r_p/r_q)^{n-1} - (1 - r_p/r_q)^{n-1} \right]$$

$$\sim 2 M \, r_q^{n-1} \left(r_p / r_q \right)$$

$$\sim K \, w (b (0{:}x_q, \delta)) \, r_p / r_q$$

for some constant K depending only on the dimension. In view of (2.4.3) we can choose A so that for sufficiently large m, the total area of the shadows $b (0{:}x_p, \delta)$ of the balls $\Delta(x_p, \delta) \in I_{m+1}$ that are partially, but not totally, eclipsed by some $\Delta(x_q, \delta) \in I_m$ will be less than

$$\frac{1}{3} \sum_{\Delta(x_q, \delta) \in I_m} w (b (0{:}x_q, \delta)) \, .$$

We pass now to consider those balls in I_{m+1} which are totally eclipsed by some member of I_m. We need an auxiliary lemma.

Lemma 2.4.7   If $\xi \in b (0{:}x_q, \delta) \cap b (0{:}x_p, \delta)$ with $\Delta(x_q, \delta) \in I_m$ and $\Delta(x_p, \delta) \in I_{m+1}$ then the geodesic $x_q \xi$ intersects the ball $\Delta(x_p, 2\delta)$.

Proof. Map the unit ball conformally onto the upper half space so that the origin goes to the point $e_n = (0,0,...,1)$ and ξ goes to infinity. Figure 2.4.2 illustrates the situation. We keep the names of the points x_p, x_q. The geodesic 0ξ becomes a vertical line through e_n and $x_q \xi$ a vertical line through x_q whose intersection with R^{n-1} we denote by c. Let b_p, b_q be the closest points to x_p, x_q on the vertical through e_n and let c_p be the closest point to b_p on the vertical through c. The non-Euclidean distances are computed as follows

$$\rho(x_q, b_q) = \int_\phi^{\pi/2} \frac{d\theta}{\sin \theta} = \log \cot (\phi/2)$$

$$\rho(b_p, c_p) = \int_\psi^{\pi/2} \frac{d\theta}{\sin \theta} = \log \cot (\psi/2) \, .$$

But

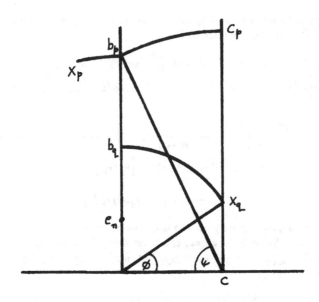

Figure 2.4.2

$$\cos\phi \;=\; \frac{|c|}{|x_q|} \;=\; \frac{|c|}{|b_q|}$$

$$\cos\psi \;=\; \frac{|c|}{|c - b_p|} \;<\; \frac{|c|}{|b_p|} \;<\; \frac{|c|}{|b_q|} = \cos\phi$$

so that $\psi > \phi$ and

$$\rho(b_p,c_p) \;<\; \rho(x_q,b_q) \;\leq\; \delta\,.$$

It follows that

$$\rho(x_p,c_p) \;\leq\; \rho(x_p,b_p) + \rho(b_p,c_p) \;\leq\; 2\delta$$

and the lemma is proved. $\square$

On the left in figure 2.4.3 we have a $\Delta(x_q,\delta)$ and some totally eclipsed $\Delta(x_p,\delta)$. On the right we have applied the map γ_q The image $\gamma_q(b(0{:}x_p,\delta))$ is not the same as $b(0{:}\gamma_q(x_p),\delta)$ but the lemma tells us that $\gamma_q(b(0{:}x_p,\delta))$ is contained in $b(0{:}\gamma_q(x_p),2\delta)$. We may use condition (2.4.1) with δ replaced by 2δ and so, for m big enough,

$$\sum w(\gamma_q(b(0{:}x_p,\delta))) \;\leq\; \sum w(b(0{:}\gamma_q(x_p),2\delta)) \;<\; \epsilon \qquad (2.4.4)$$

where the sum is taken over all x_p defining balls in I_{m+1} completely eclipsed by a

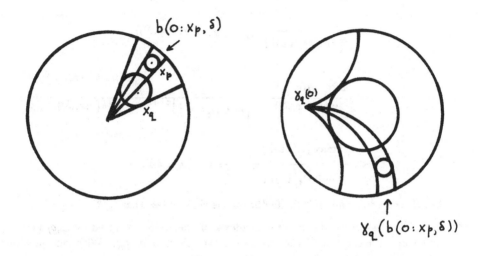

Figure 2.4.3

fixed ball of center x_q in I_m. On the other hand, if A increases to ∞ then $\gamma_q(0)$ approaches the boundary while the area of the shadow of $\Delta(0,\delta)$ viewed from $\gamma_q(0)$ decreases to a positive limit α_0. In other words,

$$w(\gamma_q(b(0{:}x_q,\delta))) \; > \; \alpha_0 \,. \tag{2.4.5}$$

But

$$w(\gamma_q(b(0{:}x_p,\delta))) \; > \; \Big[\min |\gamma_q{}'(x)|\Big]^{n-1} w(b(0{:}x_p,\delta)) \tag{2.4.6}$$

$$w(\gamma_q(b(0{:}x_q,\delta))) \; < \; \Big[\max |\gamma_q{}'(x)|\Big]^{n-1} w(b(0{:}x_q,\delta)) \tag{2.4.7}$$

where the maximum and minimum refer to $x \in b(0{:}x_q,\delta)$. Thus from (2.4.4) through (2.4.7) we see that

$$\sum_p w(b(0{:}x_p,\delta)) = \sum_p \frac{w(b(0{:}x_p,\delta))\,w(\gamma_q(b(0{:}x_p,\delta)))}{w(\gamma_q(b(0{:}x_p,\delta)))}$$

$$< \frac{1}{\Big[\min|\gamma_q{'}(x)|\Big]^{n-1}} \sum_p w(\gamma_q(b(0{:}x_p,\delta)))$$

$$= \frac{w(\gamma_q(b(0{:}x_q,\delta)))}{w(b(0{:}x_q,\delta))}\, \frac{1}{\Big[\min|\gamma_q{'}(x)|\Big]^{n-1}}\, \frac{\sum\limits_p w(\gamma_q(b(0{:}x_p,\delta)))}{w(\gamma_q(b(0{:}x_q,\delta)))}\, w(b(0{:}x_q,\delta))$$

$$< \frac{\Big[\max|\gamma_q{'}(x)|\Big]^{n-1}}{\Big[\min|\gamma_q{'}(x)|\Big]^{n-1}}\, \frac{\epsilon}{\alpha_0}\, w(b(0{:}x_q,\delta))\,.$$

We choose ϵ so small that the factor on the right is less than $1/3$.

Thus we conclude that the shadows of the totally eclipsed $\Delta(x_p,\delta)$ in I_{m+1} make up at most $1/3$ of the shadows of the $\Delta(x_q,\delta)$ in I_m. With our previous estimate for the partially eclipsed balls we have

$$\sum_{\Delta(x_p,\delta)\in I_{m+1}} w(b(0{:}x_p,\delta)) < \frac{2}{3} \sum_{\Delta(x_q,\delta)\in I_m} w(b(0{:}x_q,\delta))$$

for all sufficiently large m. This in turn proves that

$$\sum_{p=0}^{\infty} w(b(0{:}x_p,\delta)) < \infty$$

which is (2.4.2) and the proof of the theorem is complete. $\square$

For certain groups we may specify completely the nature of the conical limit set.

Theorem 2.4.8   Let Γ be a discrete group for which the closure of the Dirichlet region D is contained in B, then every point of S is a conical limit point.

Proof.   There exists $\epsilon > 0$ such that D is contained in $\{x : |x| < 1-\epsilon\}$. Now choose $\xi \in S$ and let σ denote the radius to ξ. The radius σ clearly meets infinitely many Γ-images of D and is thus a bounded hyperbolic distance from infinitely many Γ-images of 0. From theorem 1.2.4 we see that infinitely many Γ-images of 0 approach ξ within a cone. Thus ξ is a conical limit point. $\square$

Theorem 2.4.9   If Γ is a convex co-compact discrete group preserving B then every limit point of Γ is a conical limit point.

Proof. Denote by $C(\Lambda)$ the convex hull in B of the limit set Λ. Select ξ, a limit point for Γ and u, a point of $C(\Lambda)$. The geodesic ray σ from u to ξ lies entirely in $C(\Lambda)$. But $C(\Lambda)$ has a compact fundamental region, say D and now we may argue exactly as in the proof of the previous result. $\square$

In general, a conical limit point cannot appear on the boundary of a Dirichlet region.

Theorem 2.4.10   Let Γ be a discrete group preserving the unit ball B. If D denotes the Dirichlet region centered at the origin and if $\xi \in \partial D \cap S$ then ξ is not a conical limit point.

Proof. If we suppose that $\xi \in \partial D$ then for any c, $0<c<1$, $c\xi \in D$ and so the point $c\xi$ is closer to 0 than to any Γ-image of 0. Thus defining the ball

$$B_c = \{w : \rho(w,c\xi) < \rho(c\xi,0)\}$$

we see that B_c contains no Γ-images of 0. It follows that

$$\left(\bigcup_{c \in (0,1)} B_c \right) \cap \Gamma(0) = \varnothing.$$

However, the union $\bigcup_{c \in (0,1)} B_c$ is easily seen to be a ball of radius $1/2$ centered at $\xi/2$ and this ball contains cones at ξ of arbitrarily wide opening. Thus no such cone contains infinitely many images of the origin and the point ξ is not a conical limit point. $\square$

2.5 The Horospherical Limit Set

Analogous to the notion of orbits approaching the boundary in a conical region is that of an orbit approaching the boundary in a horosphere. We first give the analytic definition.

Definition. Let Γ be a discrete group acting in B. A point $\xi \in S$ is a horospherical limit point for Γ if for every $a \in B$ there exists a sequence $\{\gamma_n\} \subset \Gamma$ such that

$$\frac{|\xi - \gamma_n(a)|^2}{1 - |\gamma_n(a)|} \to 0 \quad \text{as} \quad n \to \infty.$$

We see from theorem 1.2.5 that at a horospherical limit point ξ, the orbit of every point of B enters every horoball at ξ.

The horospherical limit set is denoted by H and the following result is an immediate consequence of the definition.

Theorem 2.5.1   Let Γ be a discrete group acting in B then

$$H \; = \; \bigcap_{k>0} L(a:k,1/2).$$

In terms of derivatives, we note from theorem 1.3.4 that $\xi \in H$ if and only if $\{|\gamma'(\xi)| : \gamma \in \Gamma\}$ is an unbounded set of reals.

We next consider image horospheres.

Lemma 2.5.2   Suppose $\xi \in S$ and A is a horosphere at ξ of Euclidean radius k. If γ is a Moebius transform preserving B then $\gamma(A)$ is a horosphere at $\gamma(\xi)$ of Euclidean radius

$$\frac{k \, |\gamma'(\xi)|}{1 - k + k|\gamma'(\xi)|}.$$

Proof. A contains the point $(1-2k)\xi$ and so $\gamma(A)$ contains the point $\gamma((1-2k)\xi)$. If we let w be the Euclidean radius of $\gamma(A)$ then, by theorem 1.2.5 we need to solve the equation

$$(1 - |\gamma((1-2k)\xi)|^2) \, |\gamma((1-2k)\xi) - \gamma(\xi)|^{-2} \; = \; (1-w)/w. \qquad (2.5.1)$$

The second term on the left hand side is found to be (using theorem 1.3.4)

$$|\gamma'((1-2k)\xi)|^{-1} \, |\gamma'(\xi)|^{-1}/4k^2$$

and, again using theorem 1.3.4, we find

$$|\gamma'((1-2k)\xi)|^{-1} = \frac{1 - (1-2k)^2}{1 - |\gamma((1-2k)\xi)|^2} \, .$$

Thus equation (2.5.1) becomes

$$\frac{1-(1-2k)^2}{4k^2|\gamma'(\xi)|} = \frac{1-w}{w}$$

solving for w,

$$w = \frac{k\,|\gamma'(\xi)|}{1 - k + k|\gamma'(\xi)|}$$

as required. $\square$

It follows from lemma 2.5.2 that any horosphere at a horospherical limit point has images of radius arbitrarily close to one. If the group is of the first kind then an argument of Hedlund [Hedlund, 1936 p.537] may be applied to show that images of any such horosphere approximate any horosphere and again using lemma 2.5.2 we obtain

Theorem 2.5.3   Let Γ be a discrete group acting in B. If Γ is of the first kind then $\xi \in H$ if and only if the set $\{|\gamma'(\xi)| : \gamma \in \Gamma\}$ is dense in the positive reals.

The notion of a horosphere having images approximating every horosphere is analogous to the notion of a line transitive point and Hedlund thus described this situation (in dimension 2) by saying that the point at infinity of the horosphere was a "horocyclic transitive point " [Hedlund, 1936]. It was his work on the characterization of such points which led him to consider what we now call conical limit points. His results were later extended to the three dimensional case — see [Tuller, 1938].

In terms of the group action on S, Sullivan has shown that the horospherical limit set H is the conservative piece. We will consider these ideas further in Chapter 6. For details on this and related topics see: [Sullivan, 1981], [Ahlfors, 1981], and [Nicholls, 1983a].

2.6 The Dirichlet Set

Our last class of boundary points is defined as follows.

Definition.   For a discrete group Γ the point $\xi \in S$ is a **Dirichlet point** if for every $a \in B$ the set

$$\left[\frac{|\gamma(a) - \xi|^2}{1 - |\gamma(a)|} : \gamma \in \Gamma \right]$$

has an attained minimum.

The set of Dirichlet points is denoted by D and in order to understand the name we need two lemmas.

Lemma 2.6.1   Let σ be the hyperbolic ray connecting $a \in B$ and $\xi \in S$. Then the horoball A at ξ through a may be written

$$A = \bigcup_{y \in \sigma} \{x : \rho(x,y) < \rho(a,y)\}.$$

Proof.   We need only observe that, for any $y \in \sigma$, the sphere $\{x : \rho(x,y) = \rho(a,y)\}$ is internally tangent to the horosphere ∂A at a. $\square$

Lemma 2.6.2   Let Γ be a discrete group acting in B and $\xi \in S$. If $a \in B$ then $\xi \in \partial D_a$ if and only if the horoball at ξ through a contains no Γ-image of a.

Proof.   Suppose $\xi \in \partial D_a$ then by convexity the geodesic ray σ joining a to ξ is in D_a. It follows that for any $y \in \sigma$ the ball $\{x : \rho(x,y) < \rho(a,y)\}$ contains no Γ-images of a and the conclusion follows from lemma 2.6.1. The proof in the reverse direction is entirely similar. $\square$

From theorem 1.2.5 we see that $\xi \in D$ if and only if for every $a \in B$ there exists $\gamma \in \Gamma$ with the property that the horoball at ξ through $\gamma(a)$ contains no Γ-image of a. In view of lemma 2.6.2 we have the following result.

Theorem 2.6.3   Let Γ be a discrete group acting in B and $\xi \in S$ then ξ is a Dirichlet point for Γ if and only if, for every $a \in B$ there exists $\gamma \in \Gamma$ with $\gamma(\xi) \in \partial D_a$.

Thus the Dirichlet set is precisely the set of points of S which are represented on the boundary of every Dirichlet region. The set D of Dirichlet points has an interesting property. If $\xi \in D$ is not a fixed point, then from each orbit $\Gamma(a)$ we may select a representative which minimizes $|\gamma(a) - \xi|^2 (1 - |\gamma(a)|)^{-1}$ and it is possible to do this in such a way as to obtain a convex fundamental region for Γ. This region has a very natural interpretation as a Dirichlet region centered at ξ in a topological sense. For details of the construction see [Beardon and Nicholls, 1982] and [Nicholls, 1984].

In terms of derivatives, we note from theorems 1.2.5 and 1.3.4 that $\gamma(0)$ belongs to the horoball at ξ of Euclidean radius k if and only if

$$|(\gamma^{-1})'(\xi)| > \frac{1-k}{k}$$

and as an immediate consequence we see that if $\xi \in D$ then the sequence $\{\gamma'(\xi)\} : \gamma \in \Gamma$ accumulates only at zero. In fact Pommerenke [Pommerenke, 1976] has shown much more.

Theorem 2.6.4   (Pommerenke) Let Γ be a discrete group acting in B. For almost all (w) $\xi \in D$ the series $\sum_{\gamma \in \Gamma} |\gamma'(\xi)|^{n-1}$ converges.

Proof.   Select $a \in B$ and write $e_a = \partial D_a \cap S$. We first prove that if $\gamma \neq I$ then $e_a \cap \gamma(e_a)$ is countable. To see this, suppose $\xi \in e_a \cap \gamma(e_a)$, then the rays joining a to ξ and a to $\gamma^{-1}(\xi)$ lie in D_a. Thus the ray from $\gamma(a)$ to ξ lies in $\gamma(D_a)$ and we deduce that the hyperbolic bisector of the segment joining a to $\gamma(a)$ ends at ξ. There are only countably many such bisectors and hence countably many such ξ. It follows that the sets $\{\gamma(e_a) : \gamma \in \Gamma\}$ overlap in at most a countable set and so the sum $\sum_{\gamma \in \Gamma} w(\gamma(e_a))$ converges. Thus

$$\sum_{\gamma \in \Gamma} \int_{e_a} |\gamma'(\xi)|^{n-1} dw = \int_{e_a} \sum_{\gamma \in \Gamma} |\gamma'(\xi)|^{n-1} dw$$

converges and it follows that the series $\sum_{\gamma \in \Gamma} |\gamma'(\xi)|^{n-1}$ converges for almost every

point of e_a. Clearly then the series converges for almost every point of $\bigcup_{\gamma \in \Gamma} \gamma(e_a)$.
But this latter set includes D and the theorem is proved. $\square$

Corollary 2.6.5   Let Γ be a discrete group acting in B. If $w(D) > 0$ then Γ converges at the exponent $n-1$.

Proof. From theorem 1.3.4 we know that for any Moebius γ and $\xi \in S$

$$|(\gamma^{-1})'(\xi)| = (1 - |\gamma(0)|^2) |\xi - \gamma(0)|^{-2}.$$

So from the theorem, the series $\sum_{\gamma \in \Gamma} (1 - |\gamma(0)|)^{n-1}$ converges. $\square$

   It should be remarked that in [Pommerenke, 1976] there is an example to show that the converse of this corollary is false (at least in dimension 2).

   Contrasting D with H we observe that $D \cap H = \varnothing$ for any group Γ but there is the possibility that $D \cup H \neq S$ and we will consider this situation later. However, the following result (implicit in [Pommerenke, 1976] in two dimensions and due to Sullivan [Sullivan, 1981] in higher dimensions) shows that $D \cup H$ comprises most of S.

Theorem 2.6.6   Let Γ be a discrete group acting in B. The sphere S may be written as the disjoint union

$$S = H \cup D \cup Q$$

where $w(Q) = 0$.

Proof. Using theorem 2.5.1 and corollary 2.1.3 we observe that for any discrete group the set H has the same measure (w) as the set

$$\bigcup_{k > 0} L(a:k,1/2)$$

but this latter set comprises those points ξ of S with the property that an orbit enters **some** horoball at ξ infinitely often. The complement of this set comprises those points η of S with the property that **every** horoball at η meets every orbit finitely often. Such points, in view of lemma 2.6.2, are points of D and the theorem is proved. $\square$

   What can be said about the set Q of theorem 2.6.6 ? If $\xi \in Q$ then for some $a \in B$ the set

$$\left\{ \frac{|h(a) - \xi|^2}{1 - |h(a)|} \; : \; \gamma \in \Gamma \right\}$$

is bounded away from zero and does not have an attained minimum. Geometrically this means that there is a critical horoball based at ξ containing no Γ-equivalents of a but with the property that **any** larger horoball contains infinitely many such equivalents. Such a limit point is called a **Garnett point** and such objects are known to exist in all dimensions — see [Nicholls, 1980] for example. They can arise as limit points which are represented on the boundary of some but not all Dirichlet regions for the group.

2.7 Parabolic Fixed Points

Let us suppose that a discrete group Γ acts in the upper half space H of R^n and that ∞ is fixed by a parabolic transform in Γ. Write Γ_∞ for the stabilizer subgroup of ∞. As we have seen in section 1.3, Γ_∞ acts as a group of Euclidean isometries of $\partial H = R^{n-1}$. We make the almost trivial observation that if D is a fundamental domain for the action of Γ_∞ on R^{n-1} then $D \times (0, \infty)$ is a fundamental domain for the action of Γ_∞ on H.

We next describe some properties of Γ_∞ and refer the reader to [Bowditch, 1988] for the proofs - these results are also to be found (implicitly) in [Wolf, 1974, p.100]. The group Γ_∞ possesses a normal subgroup of finite index, say G, and there is a non empty G-invariant plane in R^{n-1} on which G acts as a group of Euclidean translations. We denote this plane by P_∞ and call it the translation plane of Γ_∞. An example might be helpful at this point.

Consider the orthogonal matrix A defined by

$$A = \begin{pmatrix} \cos\theta & -\sin\theta & 0 & 0 & 0 & 0 \\ \sin\theta & \cos\theta & 0 & 0 & 0 & 0 \\ 0 & 0 & 1 & 0 & 0 & 0 \\ 0 & 0 & 0 & 1 & 0 & 0 \\ 0 & 0 & 0 & 0 & 1 & 0 \\ 0 & 0 & 0 & 0 & 0 & 1 \end{pmatrix}$$

and the vectors $\alpha = (0,0,0,1,0,0)$ $\beta = (0,0,1,0,0,0)$. The parabolic transforms P_1, P_2, of H the upper half space in R^6, defined by

$$P_1(x) = Ax + \alpha \qquad P_2(x) = x + \beta$$

clearly generate a discrete group which could arise as the stabilizer of ∞ in some discrete Γ. Note that $< P_1, P_2 >$ acts as a group of Euclidean translations on the 4-dimensional plane

$$P_\infty = ((0,0,x_1,x_2,x_3,x_4) : x_i \in R, \ i = 1,2,3,4).$$

In general, there will be a subplane of P_∞ preserved by Γ_∞ and whose quotient by Γ_∞ is compact [Bowditch, 1988 p.13]. In the example above, this is the 2-dimensional subplane

$$((0,0,x_1,x_2,0,0) : x_i \in R, \ i = 1,2).$$

In the general case, the dimension of this subplane is defined to be the **rank** of the parabolic fixed point.

As an application of these ideas we prove a result which will be needed in the next Chapter.

Theorem 2.7.1   Let Γ be a non-elementary discrete group preserving H with ∞ a parabolic fixed point of rank k, then the number of transforms $\gamma \in \Gamma_\infty$ such that $m \leq |\gamma(x_1) - x_1| < m+1$, where $x_1 = (0,0,0,...,0,1)$, does not exceed a constant multiple of m^{k-1}.

Proof.   Note that Γ_∞ possesses a subgroup G of finite index which has an action as a group of Euclidean translations on a copy of R^k. We consider G as acting on R^k and wish to estimate the number of $\gamma \in G$ such that $m \leq \gamma(0) < m+1$. An upper bound is easily obtained by a volume argument. Place a small ball about 0 and consider the number of its images in the shell $\{x : m \leq |x| < m+1\}$. The volume of this shell is a constant times m^{k-1}. Now consider the action of G in the $n-1$-dimensional space $\{x \in R^n : x_n = 1\}$ and it follows that the number of γ in G with

$$m \leq |\gamma(x_1 - x_1| < m+1$$

is at most a constant times m^{k-1}. Noting that G is of finite index in Γ_∞, the proof of the theorem is complete. $\square$

We next introduce the notion of a bounded parabolic fixed point [Bowditch, 1988, p.14]. As before, denote by Γ_∞ the stabilizer of the parabolic fixed point ∞ and write σ_∞ for the minimal plane preserved by Γ_∞ and whose quotient by Γ_∞ is compact. We say that ∞ is a **bounded parabolic fixed point** if the Euclidean distance between y and σ_∞ remains bounded for all limit points y (except ∞). It is not difficult to see that this happens if and only if the quotient by Γ_∞ of the limit set minus ∞ is compact.

When $n = 3$ this notion was introduced in [Beardon and Maskit, 1974] — they called such a parabolic fixed point *cusped*. They proved that a group is geometrically finite if and only if the limit set comprises cusped parabolic fixed points and conical limit points. In higher dimensions we could adopt this property of the limit set as the definition of geometrical finiteness — it is equivalent to several other natural notions of geometrical finiteness (see [Bowditch, 1988] for a full account). However, we stay with the definition given in Chapter 1 — namely, that a group is geometrically finite if it possesses some convex fundamental polyhedron with finitely many faces. This is in fact a more restrictive definition. The methods of [Beardon and Maskit, 1974] extend to higher dimensions and yield a proof of the following result.

Theorem 2.7.2   Let Γ be a discrete group preserving H. If Γ is geometrically finite then the limit set comprises bounded parabolic fixed points and conical limit points.

As an application of these ideas we conclude this section with an estimate on the size of cuspidal ends of the quotient space H/Γ near a bounded parabolic fixed point. This result will be used in Chapter 9.

Theorem 2.7.3   Let Γ be a discrete group preserving H with ∞ a bounded parabolic fixed point of rank k. If $C(\Lambda)$ is the convex hull of the limit set Λ of Γ and if $C_1(\Lambda)$ denotes a unit neighborhood of $C(\Lambda)$ then the hyperbolic cross sectional area of $C_1(\Lambda)/\Gamma$ at height t is $O(t^{-k})$ as $t \to \infty$.

Proof.   Since ∞ is a parabolic fixed point of rank k then we know that there is a k-dimensional minimal hyperplane σ_∞ preserved by Γ_∞ and whose quotient by Γ_∞ is compact. Without loss of generality we assume that

$$\sigma_\infty = \{(x_1, x_2, ..., x_k, 0, 0, ..., 0) ; x_i \in R, i = 1, ..., k\}.$$

Note that if $y = (y_1, y_2, ..., y_{n-1}, 0)$ is a limit point for Γ then there exists M with $|y_i| < M$, $i = k+1, ..., n-1$ — this is because ∞ is a bounded parabolic fixed point. It is immediate that $C(\Lambda)$ is contained in the region

$$\{(x_1, x_2, ..., x_k, x_{k+1}, ..., x_{n-1}, x_n) : x_n \geq 0 ; |x_i| < M , i = k+1, ..., n-1\}.$$

One easily verifies that, for some T , $C_1(\Lambda) \bigcap \{ x : x_n > T\}$ is contained in the region

$$\{(x_1, x_2, ..., x_k, x_{k+1}, ..., x_{n-1}, x_n) : x_n \geq T ; |x_i| < M + 2x_n , i = k+1, ..., n-1\}.$$

since unit hyperbolic separation at height x_n corresponds to a Euclidean separation of an amount x_n. The quotient of the region above by Γ is contained in the quotient by Γ_∞ and this in turn is contained in some region

$$\{(x_1, x_2, ..., x_n) : x_n \geq T ; |x_i| < A , i = 1, ..., k; |x_i| < M + 2x_n , i = k+1, ..., n-1\}.$$

The non-Euclidean cross sectional area of this region at height t is clearly given by $(2A)^k [2(M + 2t)]^{n-1-k} / t^{n-1}$ which is $O(t^{-k})$ as $t \to \infty$. $\square$

CHAPTER 3

A Measure on the Limit Set

3.1 Construction of an Orbital Measure

In this chapter we will follow the work of Patterson [Patterson, 1976a] and Sullivan [Sullivan, 1979] in constructing a remarkable measure which is supported on the limit set of a discrete group. In order to understand the local properties of this measure and to see that it is unique (at least for convex co-compact groups) it will be necessary to go into the more general question of conformal densities — this will be done in chapter four. In view of the critical role played by this measure in the ergodic theory, we give in this chapter full details of its construction. The results in the first four sections are due to Patterson [Patterson, 1976a] and Sullivan [Sullivan, 1979]. Our point of view and notation are somewhat different.

The hyperbolic distance between two points x,y in the unit ball B will be denoted throughout this chapter by (x,y). Let Γ be a discrete group of Moebius transforms preserving B and for $x, y \in B$, $r > 0$ recall the orbital counting function N(r,x,y) of section 1.5

$$N(r,x,y) = card\{\gamma \in \Gamma\colon (x,\gamma y) < r\}.$$

For $x,y \in B$ the quantity

$$\delta_{x,y} = \limsup_{r \to \infty} \frac{1}{r} \log N(r,x,y)$$

is finite (in fact it does not exceed $n-1$ by theorem 1.5.1). Considering the Poincaré series

$$g_s(x,y) = \sum_{\gamma \in \Gamma} e^{-s(x,\gamma y)}$$

and the partial sum

$$\sum_{\gamma \in \Gamma \,:\, (x,\gamma y) \,<\, R} e^{-s(x,\gamma y)} = N(R,x,y)e^{-sR} + s \int_0^R N(t,x,y)e^{-st}\,dt$$

we see that the series converges if $s > \delta_{x,y}$ and diverges if $s < \delta_{x,y}$. The triangle inequalities:

$$(x,\gamma y) \leq (x,y) + (y,\gamma y) \ , \ (x,\gamma y) \geq (y,\gamma y) - (x,y)$$

yield

$$e^{-s(x,y)}g_s(y,y) \leq g_s(x,y) \leq e^{s(x,y)}g_s(y,y)$$

and it follows that $\delta_{x,y}$ depends on Γ only and not on x or y. The exponent will thus be written $\delta(\Gamma)$, or simply δ. It is the critical exponent of section 1.6 and we note again that $\delta \leq n-1$.

In the construction of the measure an important distinction has to be made between those groups whose Poincaré series converges at the critical exponent and those for which the series diverges at the critical exponent. Recall from section 1.6 that the group is of **convergence type** in the former case and of **divergence type** in the latter.

For $x,y \in B$ and $s > \delta$ the basic idea is to construct a measure by placing a Dirac point mass of weight $\dfrac{e^{-s(x,\gamma y)}}{g_s(y,y)}$ at each point γy. One then appeals to Helly's theorem to obtain a measure in the limit as $s \to \delta^+$. It will readily be seen that if the group is of divergence type then all the mass of this limit measure will be swept off to the limit set (since $g_\delta(y,y) = \infty$) and the construction is complete. However, if the group is of convergence type one simply obtains a new measure with point masses on the orbit of y. To overcome this difficulty, the point masses are multiplied by a factor $h(e^{(x,\gamma y)})$ which will not alter the critical exponent of the series but which will ensure divergence at that exponent. The required properties of this function h are given in the following lemma due to Patterson [Patterson, 1976a p.245].

Lemma 3.1.1   Let Γ be a discrete group with critical exponent δ. There exists a function $h : R^+ \to R^+$ which is continuous, non-decreasing, and

1.   the series $\displaystyle\sum_{\gamma \in \Gamma} e^{-s(x,\gamma y)}h(e^{(x,\gamma y)})$ converges for $s > \delta$ and diverges for $s \leq \delta$
      and

2.   if $\epsilon > 0$ is given there is r_0 such that for $r > r_0$, $t > 1$, $h(rt) \leq t^\epsilon h(r)$.

As a consequence of property (2) above we note that for t in a bounded interval of R

$$\frac{h(e^{r+t})}{h(e^r)} \longrightarrow 1$$

uniformly as $r \longrightarrow \infty$.

Proof. (Patterson) We write $\Gamma = \{\gamma_n : n = 1,2,...\}$ ordered in such a way that $a_n = e^{(x,\gamma_n y)}$ increases to infinity. Let $\{\epsilon_n\}$ be a sequence of positive numbers decreasing to zero. We will define a sequence $\{X_n\}$, with $X_n \longrightarrow \infty$, and h on the interval $[X_n, X_{n+1}]$ inductively. Let $X_1 = 1$ and set $h(x) = 1$ on $[0,1]$.

If h is defined on $[0, X_n]$ then choose X_{n+1} so that

$$\frac{h(X_n)}{X_n^{\epsilon_n}} \sum_{X_n < a_p \leq X_{n+1}} a_p^{-(\delta - \epsilon_n)} \geq 1 . \tag{3.1.1}$$

This can always be done as $\sum a_p^{-(\delta - \epsilon_n)}$ diverges. Now if $x \in [X_n, X_{n+1}]$ define

$$h(x) = h(X_n) \left(\frac{x}{X_n}\right)^{\epsilon_n} . \tag{3.1.2}$$

It is clear that $\sum h(a_p) \cdot a_p^{-\delta}$ diverges because

$$\sum h(a_p) \cdot a_p^{-\delta} = \sum_{n=1}^{\infty} \sum_{a_p \in [X_n, X_{n+1}]} h(X_n) \cdot \left(a_p/X_n\right)^{\epsilon_n} \cdot a_p^{-\delta} \geq \sum_{n=1}^{\infty} 1$$

by (3.1.1). From (3.1.2) we note that h is positive and increasing. Given $\epsilon > 0$ find n so that $\epsilon > \epsilon_n$. If $x \geq X_n$ then $\log h(x)$ is, by (3.1.2), a piecewise continuous function of $\log x$ and the slope of each component is at most ϵ_n. Thus if $y > X_n$ and $x > 1$

$$\log h(xy) - \log h(y) \leq \epsilon \log x$$

which is just what we want.

It remains only to show that $\sum h(a_p) \cdot a_p^{-s}$ converges if $s > \delta$. Choose $\epsilon > 0$ so that $\delta + \epsilon < s$. Then, as $p \longrightarrow \infty$, $h(a_p) = O(a_p^\epsilon)$ by what we have just proved. The convergence of the series follows at once, and the lemma is proved. $\square$

Before proceeding with the construction of the measure we derive an estimate on the function h. This estimate is stated in the context of the upper half space H of R^n as this is the form in which it will later be used. For a positive real number z we write $x_z = (0,0,...,0,z)$ and we have the following.

Lemma 3.1.2   If Γ is a discrete group preserving H, if $\epsilon > 0$, and $y \in H$ then there exists a constant B depending on Γ, y, and ϵ such that for all $z > 1$ and all $V \in \Gamma$,

$$h(e^{(x_s, Vy)}) < B\, z^\epsilon\, h(e^{(x_1, Vy)}).$$

Proof. Choose $\epsilon > 0$ and $y \in H$. Note that $e^{(x_s, Vy)} \leq e^{(x_s, x_1)} \cdot e^{(x_1, Vy)}$ and that $e^{(x_s, x_1)} = z$. Applying lemma 3.1.1 we see that there exists r_0 so that if $(x_1, Vy) > r_0$ then

$$h(e^{(x_s, Vy)}) < z^\epsilon\, h(e^{(x_1, Vy)}). \tag{3.1.3}$$

Thus there exists a finite set of $V \in \Gamma$, say $\{V_1, ..., V_k\}$, with the property that (3.1.3) holds for all V not in this set and all $z > 1$. Now we have only to deal with a finite set of V and we may choose $w > 1$ so that $(x_w, V_i(y)) > r_0$ for $i = 1, .., k$. If $1 < z \leq w$ then $\{h(e^{(x_s, V_i(y))}) : i = 1, .., k\}$ is bounded above by a constant depending upon Γ and y — the conclusion of the lemma is trivially true for such z. Now suppose $z > w$ and set $u = z/w$. We have

$$h(e^{(x_s, Vy)}) \leq h(e^{((x_s, x_w) + (x_w, Vy))}) = h(u e^{(x_w, Vy)}).$$

But $u > 1$ and $(x_w, Vy) > r_0$ if $V \in \{V_1, ..., V_k\}$ and so for such V we may appeal again to lemma 3.1.1 obtaining

$$h(e^{(x_s, Vy)}) < u^\epsilon\, h(e^{(x_w, Vy)}) = z^\epsilon\, h(e^{(x_1, Vy)}) \left[\frac{h(e^{(x_w, Vy)})}{h(e^{(x_1, Vy)})\, w^\epsilon} \right].$$

However, $w > 1$ and the term in square brackets above is bounded by a constant depending on ϵ, y, and Γ if $V \in \{V_1, ..., V_k\}$. This, with (3.1.3), completes the proof. $\square$

We return now to the construction of the measure. If Γ is of divergence type we may simply take h to be 1. With this convention in mind we now modify the Poincaré series defining

$$g_s^*(y, y) = \sum_{\gamma \in \Gamma} e^{-s(y, \gamma y)} h(e^{(y, \gamma y)})$$

and form the measure

$$\mu_{x, y, s} = \frac{1}{g_s^*(y, y)} \sum_{\gamma \in \Gamma} e^{-s(x, \gamma y)} h(e^{(x, \gamma y)})\, D(\gamma y)$$

for $x, y \in B$, $s > \delta$ and with $D(\gamma y)$ denoting the Dirac point mass of weight one at γy. Thus for a Borel set E in $\bar{B}$ we have

$$\mu_{x,y,s}(E) = \frac{1}{g_s^*(y,y)} \sum_{\gamma \in \Gamma} e^{-s(x,\gamma y)} h\left(e^{(x,\gamma y)}\right) 1_E(\gamma y)$$

where 1_E is the characteristic function of E. Note once again that if Γ is of divergence type we may take

$$\mu_{x,y,s}(E) = \frac{1}{g_s(y,y)} \sum_{\gamma \in \Gamma} e^{-s(x,\gamma y)} 1_E(\gamma y).$$

We will not be changing the point y and so, with the y dependence implicit, we will henceforth write $\mu_{x,s}$.

3.2 Change in Base Point

In this section we will consider how the measure $\mu_{x,s}$ varies with x as s $(>\delta)$ is held fixed. In order to do this we will have to compare terms such as $e^{-s(x,y)}$ and $e^{-s(x',y)}$.

Lemma 3.2.1   Given x, x', w in B and $\xi \in \partial B$ then

$$\frac{e^{(x,w)}}{e^{(x',w)}} \quad \longrightarrow \quad \frac{P(x',\xi)}{P(x,\xi)} \quad as \quad w \to \xi$$

where $P(x,\xi)$ is the Poisson kernel $\dfrac{1 - |x|^2}{|x - \xi|^2}$.

Proof. From [Beardon, 1983 p.131] we note that $4 \sinh^2((x,w)/2) = |x - w|^2 (1 - |x|^2)^{-1} (1 - |w|^2)^{-1}$. The left hand side is asymptotic to $e^{(x,w)}$ as $(x,w) \to \infty$. The right hand side is asymptotic to $P(x,\xi)^{-1}(1 - |w|^2)^{-1}$ as $w \to \xi$ and the result follows immediately.□

The following estimate compares the size of $\mu_{x,s}(E)$, $\mu_{x',s}(E)$ if E is a Borel set contained in a small neighborhood of $\xi \in \partial B$.

Theorem 3.2.2   Let Γ be a discrete group with critical exponent δ and suppose $s > \delta$. Choose $x, x' \in B$ and $\xi \in \partial B$. Let E be a Borel subset of $\bar{B}$ and, for $t > 0$, let $E(t)$ be the part of E within a Euclidean distance t of ξ. For $\epsilon > 0$ there exists $t(\epsilon)$ such that if $t < t(\epsilon)$

$$\left[\left(\frac{P(x,\xi)}{P(x',\xi)} \right)^s - \epsilon \right] (1-\epsilon)\mu_{x',s}(E(t)) \leq \mu_{x,s}(E(t))$$

$$\leq \left[\left(\frac{P(x,\xi)}{P(x',\xi)} \right)^s + \epsilon \right] (1+\epsilon)\mu_{x',s}(E(t)).$$

Proof. We write $U(t)$ for that part of $\bar{B}$ within a Euclidean distance t of ξ and so $E(t)$ is contained in $U(t)$, we may thus write

$$\mu_{x,s}(E(t)) = \frac{1}{g_s^*(y,y)} \sum_{\gamma \in \Gamma : \gamma y \in U(t)} e^{-s(x,\gamma y)} h\left(e^{(x,\gamma y)}\right) 1_E(\gamma y)$$

which equals

$$\frac{1}{g_s^*(y,y)} \sum_{\gamma \in \Gamma : \gamma y \in U(t)} \frac{e^{-s(x,\gamma y)}}{e^{-s(x',\gamma y)}} \frac{h\left(e^{(x,\gamma y)}\right)}{h\left(e^{(x',\gamma y)}\right)} e^{-s(x',\gamma y)} h\left(e^{(x',\gamma y)}\right) 1_E(\gamma y).$$

Choose $\epsilon > 0$ then, from lemma 3.1.1 and lemma 3.2.1, there exists $t(\epsilon)$ such that if $t < t(\epsilon)$ and $\gamma y \in U(t)$,

$$\left| \frac{h\left(e^{(x,\gamma y)}\right)}{h\left(e^{(x',\gamma y)}\right)} - 1 \right| < \epsilon \quad \text{and} \quad \left| \frac{e^{-s(x,\gamma y)}}{e^{-s(x',\gamma y)}} - \left[\frac{P(x',\xi)}{P(x,\xi)}\right]^{-s} \right| < \epsilon.$$

These estimates in the above expression for $\mu_{x,s}(E(t))$ yield the required result. $\square$

Corollary 3.2.3 With x, x' and $E(t)$ as in the theorem

$$\lim_{t \to 0} \frac{\mu_{x,s}(E(t))}{\mu_{x',s}(E(t))} = \left(\frac{P(x,\xi)}{P(x',\xi)}\right)^s.$$

We should remark here that although it does not make sense to speak of the Radon-Nikodym derivative $\dfrac{d\mu_{x,s}}{d\mu_{x',s}}(\xi)$ — because neither measure has any mass on ∂B — the corollary above gives useful information concerning the relative sizes of $\mu_{x,s}$ and $\mu_{x',s}$ near ξ. We will see the quotient $\dfrac{P(x,\xi)}{P(x',\xi)}$ later on in the context of a genuine Radon-Nikodym derivative.

If we consider now how the measures $\mu_{x,s}$, $\mu_{x',s}$ are related when x, x' are Γ-equivalent we obtain an invariance property. For a positive finite measure μ on $\bar{B}$ and for any Moebius transform γ we define a new measure $\gamma^*\mu$ by

$$\gamma^*\mu(E) = \mu(\gamma(E)).$$

Theorem 3.2.4 Let Γ be a discrete group with critical exponent δ, $s > \delta$ and $x \in B$. Then for any $\gamma \in \Gamma$,

$$\gamma^*\mu_{x,s} = \mu_{\gamma^{-1}(x),s}.$$

Proof. Suppose V is a Moebius transform preserving B then by definition,

$$\mu_{z,s}(V(E)) = \frac{1}{g_s^*(y,y)} \sum_{\gamma \in \Gamma} e^{-s(z,\gamma y)} h(e^{(z,\gamma y)}) 1_{V(E)}(\gamma y).$$

Set $\eta = V^{-1}\gamma$ and note that $\gamma(y) \in V(E)$ if and only if $V^{-1}\gamma(y) \in E$ i.e., if and only if $\eta(y) \in E$. If $V \in \Gamma$ then as γ runs over Γ, η also runs over Γ and we have

$$\mu_{z,s}(V(E)) = \frac{1}{g_s^*(y,y)} \sum_{\eta \in \Gamma} e^{-s(V^{-1}z,\eta y)} h(e^{(V^{-1}z,\eta y)}) 1_E(\eta(y)) = \mu_{V^{-1}z,s}(E).$$

The required result follows when we replace V by γ. $\square$

3.3 Change of Exponent

In this section we consider the behavior of the measure $\mu_{z,s}$ as s approaches δ^+. We will need to use Helly's theorem for which the following will be required.

Lemma 3.3.1   Let Γ be a discrete group with critical exponent δ. For x in B the family of measures $\{\mu_{z,s} : \delta < s < \delta + 1\}$ is weakly bounded. In fact $\mu_{z,s}(\overline{B})$ is bounded independently of s in this range.

Proof. Note that

$$\mu_{z,s}(\overline{B}) = \frac{\displaystyle\sum_{\gamma \in \Gamma} e^{-s(z,\gamma y)} h(e^{(z,\gamma y)})}{\displaystyle\sum_{\gamma \in \Gamma} e^{-s(y,\gamma y)} h(e^{(y,\gamma y)})}. \tag{3.3.1}$$

By the triangle inequality $e^{-s(z,\gamma y)} \le e^{s(z,y)} e^{-s(y,\gamma y)}$, and so the numerator of (3.3.1) does not exceed

$$e^{s(z,y)} \sum_{\gamma \in \Gamma} e^{-s(y,\gamma y)} \frac{h(e^{(z,\gamma y)})}{h(e^{(y,\gamma y)})} h(e^{(y,\gamma y)}).$$

However, from lemma 3.1.1, for all except finitely many terms in the series,

$$\frac{h(e^{(z,\gamma y)})}{h(e^{(y,\gamma y)})} < 2$$

and we deduce from (3.3.1) that

$$\mu_{z,s}(\overline{B}) \le \lambda e^{s(z,y)}$$

where λ depends only on (x,y). We similarly obtain a lower bound

$$\mu_{z,s}(\overline{B}) \ge \lambda' e^{-s(z,y)}$$

and the proof is complete. $\square$

Thus for $x \in B$ we may appeal to Helly's theorem to deduce that on a sequence of values of s approaching δ^+ the measures $\mu_{z,s}$ converge (in the topology of weak convergence) to a measure μ_x on $\overline{B}$. We will write $\mu_{x,s_n} \to \mu_x$ in this case. There is no reason to suppose that a unique measure is obtained in this way — different sequences may yield different weak limits. In fact for many groups this measure **will** be unique — but we have to wait until the next chapter for results of this type. For the present then we must consider a whole family of possible limit measures and accordingly define, for $x \in B$, M_x to be the collection of measures μ_x on $\overline{B}$ with the property that for some sequence $\{s_n\}$ monotonic decreasing to δ^+, $\mu_{x,s_n} \to \mu_x$.

The measures thus constructed are the ones we are after — they are concentrated on the limit set of the group.

Theorem 3.3.2   Let Γ be a discrete group with critical exponent δ. For any $x \in B$ and any $\mu_x \in M_x$, the measure μ_x is concentrated on the limit set of Γ.

Proof. It is clearly sufficient to show that if O is a (relatively) open set in $\overline{B}$ containing no limit points then $\mu_x(O) = 0$. It is sufficient to prove that a ball Δ centered at some γy and containing no other Γ-image of y has μ_x measure 0. Since Δ is open we have, as a consequence of Helly's theorem, that

$$\liminf_{n \to \infty} \ \mu_{x,s_n}(\Delta) \geq \mu_x(\Delta)$$

where $\{s_n\}$ is the sequence, monotone decreasing to δ^+, on which $\mu_{x,s_n} \to \mu_x$. However,

$$\mu_{x,s_n}(\Delta) = \frac{e^{-s_n(x,\gamma y)}h\left(e^{(x,\gamma y)}\right)}{g_{s_n}^*(y,y)}$$

and to complete the proof of the theorem it remains only to show that

$$\lim_{n \to \infty} \ g_{s_n}^*(y,y) \ = \ \infty \ .$$

With y fixed we write $\Gamma = \{\gamma_m\}$, $m = 1,2,3,...$ so that $\lambda_m = (y,\gamma_m(y))$ is non-decreasing. We write $a_m = h\left(e^{(y,\gamma_m(y))}\right)$ and form the Dirichlet series

$$f(z) \ = \ \sum_{m=1}^{\infty} a_m \, e^{-\lambda_m z}$$

which is analytic in the half-plane $\text{Re}(z) > \delta$ [Titchmarsh, 1939 p.290]. The point $s = \delta$ is a singularity for $f(s)$ (since $a_m \geq 0$ for all m) [Titchmarsh, 1939 p.294]. Thus if s is real and monotonic decreasing to δ, $f(s)$ is unbounded as required.$\Box$

Now the measures have been constructed, we will be concerned in the next section with the relation between the classes M_x, $M_{x'}$ for $x,x' \in B$.

3.4 Variation of Base Point and Invariance Properties

Consider the collection of signed measures on ∂B endowed with the topology of weak convergence. For any $x \in B$ we may regard M_x as a topological space using the subspace topology. For $x, x' \in B$ there is a natural correspondence between M_x and $M_{x'}$ which is given in the theorem below.

Theorem 3.4.1   Let Γ be a discrete group with critical exponent δ. Choose x, x' belonging to B and for $\nu_x \in M_x$ define a new measure $\phi(\nu_x)$ by

$$\phi(\nu_x)(E) = \int_E \left[\frac{P(x',\xi)}{P(x,\xi)} \right]^\delta d\nu_x(\xi).$$

Then ϕ is a homeomorphism of M_x onto $M_{x'}$. The inverse map is given by

$$\phi^{-1}(\nu_{x'})(E) = \int_E \left[\frac{P(x,\xi)}{P(x',\xi)} \right]^\delta d\nu_{x'}(\xi).$$

Further, if ν_{x,s_j} converges weakly to ν_x as $s_j \longrightarrow \delta^+$ then ν_{x',s_j} converges weakly to $\phi(\nu_x)$.

Proof. We recall theorem 3.2.2 and, interchanging the roles of x and x', we have

$$\left[\left(\frac{P(x',\xi)}{P(x,\xi)} \right)^s - \epsilon \right] (1-\epsilon)\mu_{x,s}(E(t)) \leq \mu_{x',s}(E(t))$$

$$\leq \left[\left(\frac{P(x',\xi)}{P(x,\xi)} \right)^s + \epsilon \right] (1+\epsilon)\mu_{x,s}(E(t)).$$

Let $\{s_j\}$ be a sequence of values of s, monotonic decreasing to δ, on which μ_{x,s_j} converges weakly to ν_x. Suppose that on two subsequences $\{s_{j_k}\}$, $\{s_{j_l}\}$ the measures $\mu_{x',s}$ converge weakly to $\nu_{x'}$ and $\sigma_{x'}$ respectively. Then, from the above,

$$\left[\left(\frac{P(x',\xi)}{P(x,\xi)} \right)^\delta - \epsilon \right] (1-\epsilon)\nu_x(E(t)) \leq \nu_{x'}(E(t))$$

$$\leq \left[\left(\frac{P(x',\xi)}{P(x,\xi)} \right)^\delta + \epsilon \right] (1+\epsilon)\nu_x(E(t))$$

and the same inequalities will also hold with $\nu_{x'}(E(t))$ replaced by $\sigma_{x'}(E(t))$.

Letting $t \rightarrow 0$ we see that $\nu_{x'}$, ν_x are absolutely continuous with respect to each other and that $\sigma_{x'}$, ν_x are absolutely continuous with respect to each other, further, the Radon-Nikodym derivatives are given by

$$\frac{d\nu_{x'}}{d\nu_x}(\xi) = \frac{d\sigma_{x'}}{d\nu_x}(\xi) = \left(\frac{P(x',\xi)}{P(x,\xi)}\right)^{\delta}.$$

From this it follows that $\nu_{x'}$, $\sigma_{x'}$ are the same and that $\phi(\nu_x) = \nu_{x'}$. Thus we have proved the last statement of the theorem and have also shown that ϕ is a map from M_x into $M_{x'}$. We may clearly reverse the roles of x and x' to see that the map ψ on $M_{x'}$ given by

$$\psi(\nu_{x'})(E) = \int_E \left(\frac{P(x,\xi)}{P(x',\xi)}\right)^{\delta} d\nu_{x'}(\xi)$$

is a map from $M_{x'}$ into M_x. It follows immediately from the properties of the Radon-Nikodym derivative that ψ is the inverse of ϕ and it remains only to establish the continuity of ϕ.

Suppose $\{\nu_x^j : j = 1,2,3,...\}$ is a sequence of measures in M_x converging weakly to ν_x in M_x. Let f be continuous with compact support on ∂B then $f\left(P(x',\xi) / P(x,\xi)\right)^{\delta}$ is also continuous with compact support on ∂B and so, by weak convergence,

$$\int_{\partial B} f\left(\left(\frac{P(x',\xi)}{P(x,\xi)}\right)^{\delta}\right) d\nu_x^j \rightarrow \int_{\partial B} f\left(\left(\frac{P(x',\xi)}{P(x,\xi)}\right)^{\delta}\right) d\nu_x .$$

But this means

$$\int_{\partial B} f \; d\nu_{x'}^j \rightarrow \int_{\partial B} f \; d\nu_{x'}$$

where $\nu_{x'}^j = \phi(\nu_x^j)$ and $\nu_{x'} = \phi(\nu_x)$. It follows that $\phi(\nu_x^j)$ converges weakly to $\phi(\nu_x)$ and so ϕ is a homeomorphism. $\square$

In view of this homeomorphism we shall adopt the convention of using the same greek letter for maps in M_x and $M_{x'}$ if and only if they are equivalent under ϕ. Thus if we write $\nu_x \in M_x$ and $\nu_{x'} \in M_{x'}$ it is to be understood that $\nu_{x'} = \phi(\nu_x)$.

To recapitulate, theorem 3.4.1 says that for $\nu_x \in M_x$ and $\nu_{x'} \in M_{x'}$, ν_x and $\nu_{x'}$ are absolutely continuous with respect to each other and

$$\left(\frac{d\nu_x}{d\nu_{x'}}\right)(\xi) = \left(\frac{P(x,\xi)}{P(x',\xi)}\right)^{\delta}.$$

Alternatively, for any Borel subset A of ∂B the quantity

$$\int_A (P(x,\xi))^{-\delta} d\nu_x(\xi)$$

is independent of $x \in B$.

The connection between the measures $\nu_x, \nu_{x'}$ may also be expressed in terms of the derivative of a Moebius transform γ with $\gamma(x) = x'$ (note that we do **not** require that $\gamma \in \Gamma$). The derivative used in this connection is derived from the metric obtained on ∂B by radial projection. Start with the great circle metric d_0 on ∂B defined by

$$d_0(\xi,\eta) = |\arccos \xi.\eta|.$$

Now for $x \in B$ select a Moebius transform V preserving B such that $V(x) = 0$ and define

$$d_x(\xi,\eta) = d_0(V(\xi),V(\eta)).$$

If $x \in B$, γ is Moebius preserving B, and $\xi \in \partial B$ we define

$$|h_x'(\xi)| = \lim_{\eta \to \xi} \frac{d_x(\gamma(\xi),\gamma(\eta))}{d_x(\xi,\eta)}.$$

Lemma 3.4.2   For any Moebius transform γ preserving B, for $x \in B$ and $\xi \in \partial B$,

$$|h_x'(\xi)| = \frac{P(\gamma^{-1}(x),\xi)}{P(x,\xi)}.$$

Proof. Suppose V is a Moebius transform preserving B with $V(x) = 0$ then, by definition,

$$d_x(\xi,\eta) = d_0(V(\xi),V(\eta)).$$

If we suppose that ξ,η are close then so are $V(\xi),V(\eta)$ and we have

$$d_x(\xi,\eta) \doteq |V(\xi) - V(\eta)| = |V'(\xi)|^{1/2}|V'(\eta)|^{1/2}|\xi - \eta|$$

from (1.3.2). There is a corresponding expression for $d_x(\gamma(\xi),\gamma(\eta))$ and we have

$$|h_x'(\xi)| = \frac{|(V\gamma)'(\xi)|}{|V'(\xi)|}.$$

Now if V is Moebius with $V(x) = 0$ then we recall from section 1.3 that V can be written as T_x followed by a rotation. Thus, from (1.3.9),

$$|V'(\xi)| = |T_x'(\xi)| = \frac{1 - |x|^2}{|\xi - x|^2} = P(x,\xi).$$

From the above we see that

$$\mathsf{h}_z'(\xi)| = \frac{P(\gamma^{-1}(x),\xi)}{P(x,\xi)}$$

as required. $\square$

As a consequence of theorem 3.4.1 and lemma 3.4.2 we have

Theorem 3.4.3   If $x \in B$, γ is a Moebius transform preserving B, and E is a Borel subset of ∂B then

$$\mu_{\gamma^{-1}(x)}(E) = \int_E \mathsf{h}_z'(\xi)|^\delta d\mu_x(\xi).$$

We now turn to a consideration of the relation between $\mu_{\gamma^{-1}(x)}$ and μ_x in the case that $\gamma \in \Gamma$. Recalling the definition of $\gamma^*\mu$ the following is an immediate consequence of theorem 3.2.4.

Theorem 3.4.4   If Γ is a discrete group preserving B and if $\gamma \in \Gamma$, then for any $x \in B$

$$\gamma^*\mu_x = \mu_{\gamma^{-1}(x)}.$$

This last result, in conjunction with theorem 3.4.3, shows that for a Borel subset E of ∂B and $\gamma \in \Gamma$

$$\mu_x(\gamma(E)) = \int_E \mathsf{h}_z'(\xi)|^\delta d\mu_x(\xi).$$

Thus we have a measure behaving in essentially the same way as a δ-dimensional Hausdorff measure. This tells us that we are on the right track. The precise connection between our measure μ_x and Hausdorff δ-dimensional measure will be explored in the next chapter.

As a corollary we have the following result of Beardon [Beardon, 1968].

Corollary 3.4.5   For a non-elementary discrete group Γ the critical exponent is positive.

Proof. If the critical exponent $\delta = 0$ then for $\nu_x, \nu_{x'}$ the Radon-Nikodym derivative is identically one and thus $\nu_x = \nu_{x'}$. Theorem 3.4.4 may be applied to show that for every Borel set E of ∂B and every $\gamma \in \Gamma$, $\nu_x(E) = \nu_x(\gamma(E))$ and we have a Γ-invariant measure on the limit set. Select E in ∂B, a ball of positive ν_x measure and note that, since Γ is non-elementary, E contains distinct hyperbolic fixed points, say ξ and η. Let H_1 and H_2 be hyperbolic transforms in Γ fixing ξ and η respectively. By taking powers if necessary we may assume that $H_1(E)$ and $H_2(E)$ are both subsets of E and do not intersect. Thus

$$\nu_x(E) \geq \nu_x(H_1(E)) + \nu_x(H_2(E)) = 2\nu_x(E).$$

This contradiction completes the proof. $\square$

We summarize the properties of the measure. Any measure ν_x belonging to the class M_x of measures obtained for a discrete group Γ with critical exponent δ satisfies

- ν_x is supported on the limit set of Γ.

- For $x, x' \in B$, ν_x, $\nu_{x'}$ are absolutely continuous with respect to each other and the Radon-Nikodym derivative satisfies

$$\left(\frac{d\nu_{x'}}{d\nu_x}\right)(\xi) = \left[\frac{P(x',\xi)}{P(x,\xi)}\right]^\delta.$$

- $\gamma^* \nu_x = \nu_{\gamma^{-1}(x)}$ for $\gamma \in \Gamma$.

- $\gamma^* \nu_x = |\gamma_x'|^\delta \nu_x$.

In the following section (the last of this chapter) we consider the atomic part of the measure. At first reading, rather than disturb the flow of ideas, the reader may wish to proceed directly to chapter four. The crucial result from the next section is that the measure has no atomic part if the underlying group is geometrically finite.

3.5 The Atomic Part of the Measure

If Γ is an elementary group then the limit set is finite and consequently each measure in the class M_x is purely atomic. In this section we consider the situation for non-elementary groups. It will be shown that for large classes of groups there is no atomic part to the measure.

It will be convenient to work in the upper half space and accordingly we write

$$H = \{x \in R^n : x = (x_1, x_2, ..., x_n) \text{ and } x_n > 0\}$$

and define, for $1 \leq j \leq n$, the jth coordinate map $p_j(x) = x_j$. We use again the notation (x,y) for the hyperbolic metric obtained from the differential $\dfrac{|dx|}{p_n(x)}$ and the construction of the measure μ_x $(x \in H)$ may be carried out as before with the Poisson kernel replaced by the upper half space version

$$P(x,\infty) = p_n(x) \qquad P(x,\xi) = \frac{p_n(x)}{|x - \xi|^2} \text{ if } \xi \neq \infty.$$

Our first results characterize the stabilizer of a point mass and establish the convergence of a certain series.

Lemma 3.5.1   Suppose Γ is a non-elementary discrete group preserving H and that ∞ is a point mass for the measures M_z. If Γ_∞ — the stabilizer of ∞ — contains no parabolic elements, it is finite.

Proof.   Suppose $\gamma \in \Gamma_\infty$ then for $x \in H$, $\nu_z(\gamma^{-1}(\infty)) = \nu_z(\infty)$ and so $\nu_{\gamma(z)}(\infty) = \nu_z(\infty)$. It follows that

$$\frac{p_n(\gamma(x))}{p_n(x)} = \frac{P(\gamma(x),\infty)}{P(x,\infty)} = \left[\frac{d\nu_{\gamma(z)}}{d\nu_z}(\infty)\right]^{\frac{1}{\delta}} = 1$$

which makes sense because, from corollary 3.4.5, $\delta > 0$. Thus the elements of Γ_∞ preserve $p_n(x)$. It follows from this that if Γ_∞ contains no parabolics then it is comprised entirely of elliptic elements of finite order (loxodromic elements will not preserve $p_n(x)$). We may now appeal to the strong form of the Bieberbach theorem [Wolf, 1974 p.102] to deduce that Γ_∞ is finitely generated with a torsion free subgroup of finite index. This torsion free subgroup is necessarily the identity and the proof is complete . $\square$

Lemma 3.5.2   Suppose Γ is a non-elementary discrete group preserving H and ∞ is a point mass for the measures M_z. The sum

$$\sum p_n(V(y))^\delta,$$

over a system of coset representatives of Γ/Γ_∞, converges.

Proof.   If V_1 and V_2 are two transforms appearing in the sum then $V_1^{-1}(\infty) \neq V_2^{-1}(\infty)$ and so, since ν_z is a finite measure,

$$\sum \nu_z(V^{-1}(\infty)) < \infty \quad \text{and} \quad \sum \nu_{V(z)}(\infty) < \infty$$

from which it follows that

$$\sum \left[\frac{P(V(x),\infty)}{P(x,\infty)}\right]^\delta \nu_z(\infty) < \infty.$$

In other words $\sum p_n(V(x))^\delta < \infty$ as required. $\square$

Note from the last two lemmas that if ∞ is a point mass which is not fixed by a parabolic transform then the series

$$\sum_{\gamma \in \Gamma} p_n(V(y))^\delta$$

converges. As another consequence of the last two lemmas we show that a conical limit point can never be a point mass.

Theorem 3.5.3 If Γ is a non-elementary discrete group preserving H then a conical limit point is not a point mass for the measures M_z.

Proof. By conjugation we may suppose that the conical limit point is at infinity. Then, by definition, we may find a sequence $\{V_j\}$ of transforms in Γ and $y \in H$ such that the sequence $\{p_n(V_j(y))\}$ is strictly increasing to infinity. If ∞ is not a parabolic fixed point then by lemmas 3.5.1 and 3.5.2 it cannot be a point mass.

There remains the possibility that ∞ is both a conical limit point and a parabolic fixed point (in dimensions 2 and 3 this situation cannot occur — [Beardon and Maskit, 1974 p.5] — however, we cannot rule it out in higher dimensions). We pointed out in section 2.7 that if ∞ is a parabolic fixed point then Γ_∞ preserves any plane $\{x : p_n(x) = \lambda > 0\}$ and it follows that the sequence $\{V_j\}$ introduced above contains no two elements from the same coset of Γ/Γ_∞. If ∞ is a point mass we have a contradiction with lemma 3.5.2, and the proof of the theorem is complete. $\square$

For z a positive real number we write x_z for the point $(0,0,...,0,z)$ and our main results estimate the size of $g_s(x_z,y)$ in terms of z. As one might expect, if ∞ is a parabolic fixed point the estimate is quite different than if it is not. Crucial for our purposes is the following result of Beardon [Beardon, 1968].

Lemma 3.5.4 If Γ is a non-elementary discrete group containing a parabolic fixed point of rank k then $\delta(\Gamma)$ is strictly greater than $k/2$.

We denote by D a convex fundamental domain for Γ_∞ in its action on H. Note, from the proof of lemma 3.5.1, that D may be chosen to be of the form D $= D^* \times R^+$ where

$$D^* = \{x \in R^{n-1} : |x| < |x - \gamma(0)| \text{ all } \gamma \in \Gamma_\infty - I\}.$$

Theorem 3.5.5 Let Γ be a non-elementary discrete group preserving H and suppose that ∞ is a point mass for the measures of M_z and is also a parabolic fixed point of rank k. With $y \in D$ and notation as above, for sufficiently small $\epsilon > 0$ there exists a constant A independent of s $(\delta + 1 \geq s \geq \delta)$ and z (≥ 1) such that

$$g_s^*(x_z,y) \leq A z^{k+\epsilon-s} \sum_{V:Vy\,\in\,D} p_n(V(y))^s h(e^{(x_1,Vy)}).$$

Proof. In order to estimate the terms appearing in the series for $g_s^*(x_z,y)$ we need the following result.

Lemma 3.5.6   Let Γ be a non-elementary discrete group preserving H with ∞ a parabolic fixed point. If $\gamma \in \Gamma_\infty$ and $V(y) \in D$ then for all $z > 0$

$$e^{(\gamma(z_z), V(y))} > \frac{z^2 + |h(x_1) - x_1|^2}{4zp_n(V(y))}.$$

Proof. From [Beardon, 1983 p.130] we have

$$e^{(\gamma(z_z), V(y))} = \frac{|h(x_z) - V(y)|^2}{zp_n(V(y))} + 1 + 1 - e^{-(\gamma(z_z), V(y))}. \tag{3.5.1}$$

Writing $V(y) = u + vx_1$ with $p_n(V(y)) = v$, since $V(y) \in D$, we have from (3.5.1)

$$e^{(\gamma(z_z), V(y))} > \frac{|\gamma(0) + zx_1 - u - vx_1|^2 + zv}{zv}$$

$$= \frac{|\gamma(0) - u|^2 + (z - v)^2 + zv}{zv}$$

$$> \frac{|\gamma(0)|^2 + 4z^2 - 4zv + 4v^2}{4zv}$$

$$= \frac{|\gamma(0)|^2 + z^2}{4zv} + \frac{3z^2 - 4zv + 4v^2}{4zv} \tag{3.5.2}$$

It is easily checked that the second term on the right side of (3.5.2) is never negative and so

$$e^{(\gamma(z_z), V(y))} > \frac{|\gamma(0)|^2 + z^2}{4zv} = \frac{z^2 + |\gamma(x_1) - x_1|^2}{4zp_n(V(y))}$$

as required. $\square$

Returning to the proof of the theorem, we estimate

$$g_s^*(z_z, y) = \sum_{V \in \Gamma} e^{-s(z_z, Vy)} h(e^{(z_z, Vy)}).$$

With $\epsilon > 0$, we appeal to lemma 3.1.2 and find a constant B, depending only on Γ, y, and ϵ such that

$$h(e^{(z_z, Vy)}) < B z^{\epsilon/3} h(e^{(z_1, Vy)})$$

for all $V \in \Gamma$ and $z > 0$. Thus

$$g_s^*(x_z, y) < B z^{\epsilon/3} \sum_{V \in \Gamma} e^{-s(x_z, Vy)} h\left(e^{(x_z, Vy)}\right).$$

Writing this as a double sum over cosets

$$g_s^*(x_z, y) < B z^{\epsilon/3} \sum_{V: Vy \in D} \sum_{\gamma \in \Gamma_\infty} e^{-s(\gamma x_z, Vy)} h\left(e^{(\gamma x_z, Vy)}\right). \qquad (3.5.3)$$

For an upper bound on $h\left(e^{(\gamma x_1, Vy)}\right)$ note that $(\gamma x_1, Vy) \le (\gamma x_1, x_1) + (x_1, Vy)$ thus

$$e^{(\gamma x_1, Vy)} \le e^{(\gamma x_1, x_1)} \cdot e^{(x_1, Vy)} < C(\Gamma) |\gamma(x_1) - x_1|^2 e^{(x_1, Vy)}$$

where the estimate $e^{(\gamma(x_1), x_1)} < C(\Gamma) |\gamma(x_1) - x_1|^2$ follows directly from [Beardon, 1983 p.130]. Thus for all except finitely many $V \in \Gamma$

$$h\left(e^{(\gamma x_1, Vy)}\right) < D(\Gamma) |\gamma(x_1) - x_1|^{2\epsilon/3} h\left(e^{(x_1, Vy)}\right),$$

where we have used lemma 3.1.1 with ϵ replaced by $\epsilon/3$. With this result and lemma 3.5.6, there is a constant K independent of s $(\delta \le s \le \delta + 1)$ and z $(z > 1)$ for which the inner sum in (3.5.3) does not exceed

$$K \sum_{\gamma \in \Gamma_\infty} z^s p_n(Vy)^s (z^2 + |\gamma(x_1) - x_1|^2)^{-s} |\gamma(x_1) - x_1|^{2\epsilon/3} h\left(e^{(x_1, Vy)}\right).$$

This quantity may be rewritten

$$K p_n(Vy)^s h\left(e^{(x_1, Vy)}\right) z^{-s} \sum_{\gamma \in \Gamma_\infty} \frac{|\gamma(x_1) - x_1|^{2\epsilon/3}}{(1 + z^{-2}|\gamma(x_1) - x_1|^2)^s}, \qquad (3.5.4)$$

and the sum occurring here may in turn be rewritten

$$\sum_{m=0}^{\infty} \sum_{\substack{\gamma \in \Gamma_\infty \\ m \le |\gamma(x_1) - x_1| < m+1}} \frac{|\gamma(x_1) - x_1|^{2\epsilon/3}}{(1 + z^{-2}|\gamma(x_1) - x_1|^2)^s}.$$

This does not exceed a constant multiple of

$$\sum_{m=0}^{\infty} m^{k-1} \frac{m^{2\epsilon/3}}{(1 + m^2 z^{-2})^s}$$

by theorem 2.7.1. We bound this sum by an integral

$$\int_R t^{k-1} \frac{t^{2\epsilon/3}}{(1 + t^2 z^{-2})^s} dt$$

and change variables by $u = \dfrac{t}{z}$ to obtain

$$z^{k+2\epsilon/3}\int_R \frac{u^{k+2\epsilon/3-1}}{(1+u^2)^s}du.$$

Since $\delta > k/2$ (lemma 3.5.4) then for $\epsilon < 3/2(2\delta - k)$ the integral above converges for all $s \geq \delta$. With this information in (3.5.4) we find a constant $A(\epsilon)$ independent of s ($\delta \leq s \leq \delta+1$) and z (≥ 1) so that

$$g_s^*(x_z,y) \leq Az^{k+\epsilon-s}\sum_{V:Vy\,\in\,D}p_n(Vy)^s h(e^{(x_1,Vy)}),$$

and the proof of theorem 3.5.5. is complete. $\square$

A similar result may be obtained when ∞ is not a parabolic fixed point.

Theorem 3.5.7   Let Γ be a non-elementary discrete group preserving H and suppose that ∞ is a point mass for the measures of M_z which is not fixed by any parabolic element of Γ. For sufficiently small $\epsilon > 0$ there exists a constant A independent of s ($\delta \leq s \leq \delta+1$) such that for $z > 1$

$$g_s^*(x_z,y) \leq Az^{\epsilon-s}\sum_{V\,\in\,\Gamma}p_n(V(y))^s h(e^{(x_1,Vy)}).$$

Proof. Using estimates employed in the previous proof, we have

$$g_s^*(x_z,y) = \sum_{V\,\in\,\Gamma}e^{-s(x_z,Vy)}h(e^{(x_z,Vy)}) < Bz^{\epsilon/2}\sum_{V\,\in\,\Gamma}e^{-s(x_z,Vy)}h(e^{(x_1,Vy)})$$

where B depends only on Γ and y. From [Beardon, 1983 p.130] we have, as in the proof of lemma 3.5.6, that

$$e^{(x_z,Vy)} > \frac{|x_z - Vy|^2 + zp_n(Vy)}{zp_n(Vy)}$$

and so from the above

$$g_s^*(x_z,y) < Bz^{\epsilon/2}\sum_{V\,\in\,\Gamma}z^s p_n(Vy)^s(|x_z - Vy|^2 + zp_n(Vy))^{-s}h(e^{(x_1,Vy)}).$$

Writing $Vy = u + vx_1$ (with $p_n(Vy) = v$), and assuming $\epsilon < 4\delta$, we choose an integer p satisfying $1/p < \epsilon/2 < 2\delta$ and partition the sum above as follows

$$\sum_{m\,=\,0}^{\infty}\sum_{\substack{V\,\in\,\Gamma\\ m^p<|k|<(m+1)^p}}z^s p_n(Vy)^s(|x_z - Vy|^2 + zp_n(Vy))^{-s}h(e^{(x_1,Vy)}).$$

Now

$$|x_z - Vy|^2 + zp_n(Vy) = |x_z - u - p_n(Vy)x_1|^2 + zp_n(Vy)$$

$$= (z - p_n(Vy))^2 + |u|^2 + zp_n(Vy)$$

and so

$$2(|x_z - Vy|^2 + zp_n(Vy)) = z^2 + |u|^2 + z^2 + |u|^2 + 2(p_n(Vy) - z/2)^2 - z^2/2$$

$$> z^2 + |u|^2.$$

From this it follows that

$$g_s^*(x_z, y) < Bz^{\epsilon/2} 2^s \sum_{m=0}^{\infty} \sum_{\substack{V \in \Gamma \\ m^p \leq |k| < (m+1)^p}} z^s p_n(Vy)^s h(e^{(z_1, Vy)}) (z^2 + m^{2p})^{-s}$$

$$= Bz^{\epsilon/2} 2^s z^{-s} \sum_{m=0}^{\infty} \frac{1}{(1 + z^{-2} m^{2p})^s} \sum_{\substack{V \in \Gamma \\ m^p \leq |k| < (m+1)^p}} p_n(Vy)^s h(e^{(z_1, Vy)})$$

$$< B2^s z^{\epsilon/2-s} \sum_{V \in \Gamma} p_n(Vy)^s h(e^{(z_1, Vy)}) \sum_{m=0}^{\infty} \frac{1}{(1 + z^{-2} m^{2p})^s}.$$

The second sum occurring here is bounded by an integral

$$\int_R \frac{1}{(1 + z^{-2} t^{2p})^s} dt$$

which, with the substitution, $uz = t^p$, is equal to

$$\frac{z^{1/p}}{p} \int_R \frac{1}{(1 + u^2)^s} u^{\frac{1-p}{p}} du.$$

The integral converges since $2s > 1/p$, and we obtain

$$g_s^*(x_z, y) < Az^{\epsilon-s} \sum_{V \in \Gamma} p_n(Vy)^s h(e^{(z_1, Vy)}).$$

as required. □

We conclude the chapter with several consequences of theorems 3.5.5 and 3.5.7.

Theorem 3.5.8   If Γ is a non-elementary discrete group which diverges at its critical exponent then the measures M_z have no atomic part.

Proof. Suppose there is an atom at a parabolic vertex of rank k. By conjugation we may as well suppose that Γ acts in the upper half-space H and that this parabolic vertex is at ∞. Since Γ is of divergence type, the function h used in the construction of the measure class M_x is identically 1 and so we obtain from theorem 3.5.5 that

$$g_s(x_z, y) < Az^{k+\epsilon-s} \sum_{V:Vy \in D} p_n(Vy)^s.$$

However, this sum is over a system of coset representatives of Γ/Γ_∞ — so by lemma 3.5.2 there is a constant B independent of s ($\delta \le s \le \delta + 1$) and z ($\ge 1$) such that

$$g_s(x_z, y) < Bz^{k+\epsilon-s}.$$

But the right hand side is bounded as $s \to \delta$, and this clearly contradicts the fact that Γ is of divergence type.

If an atom were to occur at a point not fixed by a parabolic element we would proceed in exactly the same fashion using theorem 3.5.7 to deduce that $g_s(x_z, y) < Bz^{\epsilon-s}$ — once again we have a contradiction with the fact that the group is of divergence type, and the theorem is proved. $\square$

Recall the definition of a bounded parabolic fixed point from section 2.7. We prove the following.

Theorem 3.5.9   If Γ is a non-elementary discrete group and ξ is a bounded parabolic fixed point, then ξ is not a point mass for the measures M_x.

Proof. Conjugate so that Γ preserves the upper half space and the bounded parabolic fixed point is at infinity. Choose y in the upper half space and we first show that, with D a convex fundamental domain for Γ_∞, the set $\{V(y) : V \in \Gamma\} \cap D$ is bounded. Since Γ is non-elementary we may select two limit points α, β neither of which is a parabolic fixed point. Choose w on the geodesic joining α to β and note that if the set above is unbounded then there exists a sequence $\{V_n\} \subset \Gamma$ such that $V_n(w) \to \infty$ and $V_n(y) \in D$. Clearly, either $V_n(\alpha) \to \infty$ or $V_n(\beta) \to \infty$ and all members of these sequences are limit points. This clearly contradicts the definition of a bounded parabolic fixed point.

With the notation of theorem 3.5.5 we have

$$g_s^*(x_z, y) \le Az^{k+\epsilon-s} \sum_{V:Vy \in D} p_n(Vy)^s h(e^{(x_1, Vy)}).$$

where k is the rank of the parabolic fixed point at ∞. From our work above we know that

$$|x_1 - Vy| < M$$

for all V with $Vy \in D$. Thus

$$g_s^*(x_z,y) \leq AM^{2s} z^{k+\epsilon-s} \sum_{V:Vy \in D} p_n(Vy)^s |x_1 - Vy|^{-2s} h(e^{(x_1,Vy)}) .$$

But the right side is bounded by

$$B z^{k+\epsilon-s} \sum_{V:Vy \in D} e^{-s(x_1,Vy)} h(e^{(x_1,Vy)})$$

for a constant B depending only on Γ and y, and so

$$g_s^*(x_z,y) < B z^{k+\epsilon-s} g_s^*(x_1,y).$$

From this we have

$$\mu_{x_z}(\overline{H}) < B z^{k+\epsilon-\delta} \mu_{x_1}(\overline{H})$$

and hence

$$\mu_{x_z}(\infty) < C z^{k+\epsilon-\delta} \qquad\qquad (3.5.5)$$

valid for all $z > 1$ and ϵ sufficiently small. But, since ∞ is a point mass we will have

$$\mu_{x_z}(\infty) = \frac{d\mu_{x_z}}{d\mu_{x_1}}(\infty) \mu_{x_1}(\infty) = \left[\frac{P(x_z,\infty)}{P(x_1,\infty)}\right]^\delta \mu_{x_1}(\infty) = z^\delta \mu_{x_1}(\infty).$$

This, in conjunction with (3.5.5), leads to $\delta \leq k + \epsilon - \delta$ and hence to $2\delta \leq k$ — but this contradicts lemma 3.5.4, and the theorem is proved. $\square$

Recalling that a discrete group is geometrically finite if it possesses a finite sided fundamental polyhedron, we note that the next result is an immediate consequence of theorems 2.7.2, 3.5.3, and 3.5.9.

Theorem 3.5.10   (Sullivan) If Γ is a non-elementary, geometrically finite group, then the measures M_x have no atomic part.

The next two results are immediate corollaries of theorem 3.5.9.

Corollary 3.5.11   Let Γ be a discrete group acting in the unit ball B of R^n. If $\xi \in \partial B$ is a parabolic fixed point of rank $k = n-1$ then ξ is not an atom for the measure class M_x.

Corollary 3.5.12   If Γ is a Fuchsian group acting in the unit ball B of R^2 then no parabolic fixed point is an atom for the measure class M_x.

We next consider the possibility of atoms occurring at non-fixed points in the case that the group is of convergence type. Note from theorem 1.2.5 and the proof of lemma 3.5.2 (interpreted in the ball) that if μ_x has an atom at ξ and if K_γ denotes the Euclidean radius of the horosphere at ξ containing $\gamma(0)$ then

$$\sum_{\gamma \in \Gamma} \left[\frac{1 - K_\gamma}{K_\gamma} \right]^\delta < \infty .$$

Using lemma 2.5.2, this is equivalent to saying that $\sum_{\gamma \in \Gamma} |\gamma'(\xi)|^\delta < \infty$. Such a point ξ is certainly a Dirichlet point (see section 2.6), but more than this is true. In particular, since the K_γ introduced above can accumulate only at 1, any horoball at ξ meets any orbit finitely often (Pommerenke [Pommerenke, 1976] called such points "orispherical limit points"). We have proved

Theorem 3.5.13   Let Γ be a non-elementary discrete group preserving B. If the measure μ_x has an atom at ξ and ξ is not a parabolic fixed point, then any horoball at ξ meets any orbit finitely often.

As an application of some of the ideas in this section we conclude with the following result.

Theorem 3.5.14   Let Γ be a discrete group acting in the unit ball B and let A be a Borel subset of ∂B such that for some, and hence every, $x \in B$, $\mu_x(A) > 0$. If, further, $\mu_x(\gamma(A) \bigcap A) = 0$ for every $\gamma \in \Gamma$, then the group converges at its critical exponent.

Proof. Since μ_0 is a finite measure we must have

$$\sum_{\gamma \in \Gamma} \mu_0(\gamma(A)) < \infty$$

but,

$$\mu_0(\gamma(A)) = \mu_{\gamma^{-1}(0)}(A) = \int_A d\mu_{\gamma^{-1}(0)}(\xi) = \int_A \left(\frac{d\mu_{\gamma^{-1}(0)}}{d\mu_0} \right)(\xi) d\mu_0(\xi)$$

$$= \int_A P(\gamma^{-1}(0),\xi)^\delta d\mu_0(\xi)$$

and we see that

$$\sum_{\gamma \in \Gamma} \int_A P(\gamma(0),\xi)^\delta d\mu_0(\xi) < \infty.$$

Interchanging the order of summation and integration, the series

$$\sum_{\gamma \in \Gamma} P(\gamma(0),\xi)^\delta$$

must converge for almost every $\xi \in A$ (with respect to the measure μ_0). Since

$\mu_0(A) > 0$ we deduce that

$$\sum_{\gamma \in \Gamma} (1 - |\gamma(0)|^2)^\delta < \infty$$

which is the required result. $\square$

We have the following corollary of theorem 3.5.14.

Corollary 3.5.15   Let Γ be a non-elementary discrete group acting in the unit ball with Dirichlet region D_a centered at $a \in B$. We write $e_a = \overline{D}_a \cap \partial B$ and suppose that, for some $x \in B$, $\mu_x(e_a) > 0$, then Γ converges at its critical exponent.

Proof.   Suppose, under the hypothesis of the corollary, that Γ diverges at the critical exponent. Then by theorem 3.5.8 the measure μ_x has no atoms. We have seen in the proof of theorem 2.6.4 that the set e_a meets Γ-images of itself in a countable set and thus $\mu_x(e_a \cap \gamma(e_a)) = 0$ for every $\gamma \in \Gamma$. From theorem 3.5.14 we have convergence at the critical exponent and this contradiction completes the proof. $\square$

CHAPTER 4

Conformal Densities

4.1 Introduction

Recall from section 3.4 the metric d_z defined on ∂B. This metric arises from the Riemannian metric tensor $G_z(\xi) = P(x,\xi)^2 I$ defined on ∂B (which is regarded as a smooth manifold of dimension n-1). For any $x, x' \in B$

$$G_{z'}(\xi) = \left[\frac{P(x',\xi)}{P(x,\xi)} \right]^2 G_z(\xi)$$

and the metrics d_z and $d_{z'}$ are said to be **conformally equivalent**.

For a discrete group Γ acting in the unit ball B we make the following definition.

Definition A Γ-invariant conformal density of dimension α is a map σ from the collection of all metrics $\{G_z : x \in B\}$ into the collection of positive finite measures on ∂B such that, writing $\sigma_z = \sigma(G_z)$,

1. σ_z is supported on the limit set of Γ

2. for $x, x' \in B$, $\sigma_z, \sigma_{z'}$ are absolutely continuous with respect to each other and the Radon-Nikodym derivative satisfies

$$\left(\frac{d\sigma_{z'}}{d\sigma_z} \right)(\xi) = \left[\frac{P(x',\xi)}{P(x,\xi)} \right]^{\alpha}$$

3. $\gamma^* \sigma_z = \sigma_{\gamma^{-1}(z)}$ for $\gamma \in \Gamma$.

The following result is an immediate consequence of these defining properties (see the remark following theorem 3.4.4).

Theorem 4.1.1   Let Γ be a discrete group preserving B and σ a Γ-invariant conformal density of dimension α, then for E a Borel subset of ∂B and $\gamma \in \Gamma$,

$$\sigma_x(\gamma(E)) = \int_E |\gamma_x{}'(\xi)|^\alpha \, d\sigma_x(\xi).$$

The definition given above should look familiar — it is in fact the list of properties satisfied by measures ν_x belonging to the class M_x which was given at the end of section 3.4. Defining a map ν from metrics to measures by $\nu(G(x)) = \nu_x \in M_x$ (with the standard convention that for $x, x' \in B$, $\nu_x, \nu_{x'}$ are corresponding members of the two classes $M_x, M_{x'}$) we have the following result.

Theorem 4.1.2   Let Γ be a discrete group with critical exponent δ then the map ν defined above is a Γ-invariant conformal density of dimension δ.

One of our major concerns is with the existence and uniqueness of Γ-invariant conformal densities. Specifically, we would like answers to the following questions:

- Given Γ, for what positive real numbers α does a Γ-invariant conformal density of dimension α exist?

- If a Γ-invariant conformal density of dimension α exists, is there another conformal density of the same dimension which is not merely a multiple of the first?

Complete answers to these questions are not known although the work of Sullivan [Sullivan, 1984] does give answers for a large class of discrete groups. In particular, we will see in section 4.6 that, for convex co-compact discrete groups, the map ν defined above is the only Γ-invariant conformal density of **any** dimension. The importance of obtaining answers to these questions lies in the fact that the existence of a conformal density has implications for the Hausdorff dimension of the limit set of a discrete group, and that the uniqueness of conformal densities has implications for the ergodic properties of the group. In this chapter we will be following the work of Sullivan [Sullivan, 1984]. The results presented here are his, our exposition is somewhat different.

In section 4.2 we establish a condition which is necessary and sufficient for a conformal density of dimension α — if one exists — to be the unique conformal density of that dimension. In section 4.3 we make a detailed study of the local behavior of the measure class associated with a conformal density. In particular, we show that the measure behaves roughly like an α-dimensional Hausdorff measure on the conical limit set. More results on the behavior of the measure on the conical limit set are given in section 4.4 — for example it is shown that Γ always acts ergodically on the conical limit set with respect to the measure class of a conformal density. In section 4.5 we consider the orbital counting function of a discrete group and obtain an upper bound analogous to the estimate given in theorem 1.5.1. It is further shown that this bound has implications for the existence of conformal densities of various dimensions. A study of conformal densities for convex co-compact groups is given in section 4.6, and the main results of the chapter are summarized in section 4.7.

4.2 Uniqueness

Let Γ be a discrete group acting in B and suppose that for some $\alpha > 0$ a Γ-invariant conformal density of dimension α exists. We denote this density by σ. We say that Γ is **ergodic** on ∂B with respect to the measure class defined by σ if whenever a Borel set $A \subset \partial B$ is invariant under Γ then for one (and hence every) $x \in B$ either $\sigma_x(A) = 0$ or $\sigma_x(\sim A) = 0$.

In this section we prove the following result.

Theorem 4.2.1   Let Γ be a discrete group acting in the unit ball B and, for $\alpha > 0$, suppose σ is a Γ-invariant conformal density of dimension α. The collection of all Γ-invariant conformal densities of dimension α is equal to the set of non-zero constant multiples of σ if and only if Γ is ergodic on ∂B with respect to the measure class defined by σ.

Proof. Suppose first that Γ is not ergodic on ∂B with respect to the measure class defined by σ, then there exists a Borel subset A of ∂B such that A is Γ-invariant, and for all $x \in B$, $\sigma_x(A) > 0$ and $\sigma_x(\sim A) > 0$. We note from property 3 of a conformal density that $\sigma_x(A) = \sigma_{\gamma(x)}(A)$ for all $\gamma \in \Gamma$. Define a new conformal density π by

$$\pi(G_x) = \pi_x = \sigma_x|_A .$$

In other words, for any Borel set E, $\pi_x(E) = \sigma_x(A \cap E)$. Note that the measure π_x is concentrated on the limit set of Γ (since σ_x is so concentrated). Select $x, x' \in B$ and suppose $\pi_x(E) = 0$, then $\sigma_x(A \cap E) = 0$ and so $\sigma_{x'}(A \cap E) = 0$ because $\sigma_x, \sigma_{x'}$ are absolutely continuous with respect to each other. It follows that $\pi_{x'}(E) = 0$ and we see that $\pi_x, \pi_{x'}$ are absolutely continuous with respect to each other. We next compute the Radon-Nikodym derivative $\dfrac{d\pi_{x'}}{d\pi_x}(\xi)$. In order to do this we note that for any $x \in B$ the two measures π_x, σ_x on the space $(A, B(A))$ (where $B(A)$ is the collection of Borel subsets of A) are absolutely continuous with respect to each other and the associated Radon-Nikodym derivative is equal to one. Thus, for any Borel subset E of ∂B

$$\pi_{x'}(E) = \sigma_{x'}(A \cap E) = \int_{A \cap E} d\sigma_{x'}(\xi) = \int_{A \cap E} \left(\frac{d\sigma_{x'}}{d\sigma_x} \right)(\xi) d\sigma_x(\xi) .$$

This latter integral may be written in the form

$$\int_{A \cap E} \left[\frac{P(x',\xi)}{P(x,\xi)} \right]^\alpha d\sigma_z(\xi) = \int_{A \cap E} \left[\frac{P(x',\xi)}{P(x,\xi)} \right]^\alpha d\pi_z(\xi) = \int_E \left[\frac{P(x',\xi)}{P(x,\xi)} \right]^\alpha d\pi_z(\xi).$$

Thus $\dfrac{d\pi_{z'}}{d\pi_z}(\xi) = \left[\dfrac{P(x',\xi)}{P(x,\xi)} \right]^\alpha$ and the map π satisfies the first two properties of a conformal density of dimension α. To check the third property we merely note that if $\gamma \in \Gamma$ and E is a Borel subset of ∂B,

$$\pi_z(\gamma(E)) = \sigma_z(\gamma(E) \cap A) = \sigma_z[\gamma(E \cap A)] = \sigma_{\gamma^{-1}(z)}(E \cap A) = \pi_{\gamma^{-1}(z)}(E)$$

and so $\gamma^* \pi_z = \pi_{\gamma^{-1}(z)}$, for any $\gamma \in \Gamma$ and $x \in B$. Thus π is a Γ-invariant conformal density of dimension α.

For the converse we suppose that Γ is ergodic on ∂B with respect to the measure class of σ. Let π be another conformal density of dimension α and note that $\nu = (\sigma + \pi)/2$ is also a conformal density of dimension α. The measures σ_z and π_z are both absolutely continuous with respect to ν_z and so the Radon-Nikodym derivatives $\dfrac{d\sigma_z}{d\nu_z}$, $\dfrac{d\pi_z}{d\nu_z}$ exist and are measurable (ν_z). They are clearly also both measurable (σ_z). We next show that, as functions on ∂B, these two derivatives are Γ-invariant.

Choose $\xi \in \partial B$ and, for $t > 0$, let $B(\xi,t)$ be the ball in ∂B of Euclidean radius t centered at ξ. From theorem 4.1.1

$$\frac{\sigma_z[\gamma(B(\xi,t))]}{\nu_z[\gamma(B(\xi,t))]} = \frac{\int_{\Delta(t)} |\gamma_z'(\eta)|^\alpha d\sigma_z(\eta)}{\int_{\Delta(t)} |\gamma_z'(\eta)|^\alpha d\nu_z(\eta)}$$

and, using the continuity of $|\gamma_z'(\eta)|$, we may take the limit as $t \to 0$ to obtain

$$\left(\frac{d\sigma_z}{d\nu_z} \right)(\gamma(\xi)) = \left(\frac{d\sigma_z}{d\nu_z} \right)(\xi).$$

Now $d\sigma_z / d\nu_z$ is measurable (σ_z) and Γ-invariant. But the action of Γ is ergodic on ∂B with respect to σ_z and this clearly implies that the derivative is equal to a non zero constant almost everywhere (σ_z). Similar comments apply to the derivative $d\pi_z / d\nu_z$ and so, from the properties of the Radon-Nikodym derivative, $d\pi_z / d\sigma_z$ is equal almost everywhere (σ_z) to a positive constant. This completes the proof of the theorem. $\square$

4.3 Local Properties

For $x \in B$ and $c > 0$ denote by $\Delta(x,c)$ the hyperbolic ball centered at x and of radius c. If y is not a point of $\Delta(x,c)$ then denote by $b(y:x,c)$ the projection of $\Delta(x,c)$ onto ∂B from y. Thus $\xi \in b(y:x,c)$ if and only if $\xi \in \partial B$ and the geodesic from y to ξ intersects $\Delta(x,c)$. Note that this generalizes the notion of shadows introduced in section 1.2 and used in section 2.4.

Lemma 4.3.1   With notation as above, the set $b(y{:}x,c)$ is a ball in the d_y metric on ∂B whose radius, r, is given by

$$\tan r = \frac{\tanh c\ (1 - |x|^2)(1 - |y|^2)}{2\,|y|\,|x - y|\,|x - y^*|}$$

(where y^* is the reflection of y in the unit ball), and whose center is the projection of the point x on ∂B from y.

Proof. A conjugation by a Moebius transformation taking y to 0 shows that $b(y{:}x,c)$ is a ball in the d_y metric centered at the projection of x from y. To compute the radius in this metric we make the observation that r is the angle at y between the geodesic through y and x and any geodesic from y tangent to $\partial\Delta(x,c)$.

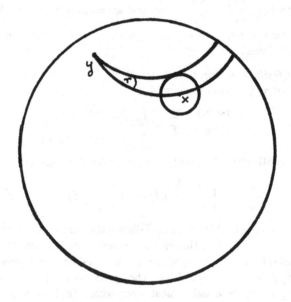

Figure 4.3.1

Let the point of tangency of such a geodesic be u, and we have a right angled hyperbolic triangle. We may apply hyperbolic trigonometry to obtain [Beardon, 1983 p.147]

$$\tan r = \frac{\tanh c}{\sinh (y,x)}.$$

The lemma now follows from the well known expressions for $\sinh((y,x)/2)$ and $\cosh((y,x)/2)$ [Beardon, 1983 p.131/2]. □

We should remark that the formula given in lemma 4.3.1 is not valid for $y = 0$ (for $0^* = \infty$). Taking a limit as $y \to 0$ we see in this case that

$$\tan r = \frac{\tanh c (1 - |x|^2)}{2|x|}.$$

Denoting the radius r by $r(y:x,c)$ we have

$$\tan r (0:x,c) = \frac{\tanh c (1 - |x|^2)}{2|x|} \qquad (4.3.1)$$

The following result gives much useful information on the local structure of a conformal density.

Theorem 4.3.2   Let Γ be a discrete group acting in B and σ a Γ-invariant conformal density of dimension α which is not a single atom. Select $x \in B$, then there exist positive constants a, A such that, provided c is large enough, for all except finitely many $\gamma \in \Gamma$,

$$a < \frac{\sigma_x[b(x:\gamma(x),c)]}{[r(x:\gamma(x),c)]^\alpha} < A.$$

Proof. Let $\lambda = (\sigma_x(\partial B) + \sigma_x(\eta))/2$ where $\sigma_x(\eta)$ is the largest point mass for σ_x (obviously we take $\sigma_x(\eta) = 0$ if σ_x has no atomic part). Since σ_x is not a single atom we have $\lambda > \sigma_x(\eta)$. Suppose on a sequence $\{\epsilon_n\}$ tending to zero we have balls in ∂B of radius ϵ_n and of σ_x mass at least λ then, on a subsequence if necessary, the centers of these balls converge to a point ξ with $\sigma_x(\xi) > \lambda$. This contradiction shows that there exists $\epsilon > 0$ such that if Δ is a ball in ∂B of d_0 radius at most ϵ then $\sigma_x(\Delta) \leq \lambda < \sigma_x(\partial B)$.

In proving the theorem we may as well suppose that $x = 0$. Now choose c so large that if $(z,0) > c$ then the set $\partial B - b(z:0,c)$ is contained in a ball of d_0 radius equal to ϵ. For the remainder of the proof c is fixed at this value. If $\gamma \in \Gamma$ with $(\gamma(0),0) > c$ (this will be true for all but finitely many $\gamma \in \Gamma$), we set $\delta = \sigma_0[b(\gamma(0):0,c)]$ and note from our remarks above that

$$\delta \geq \sigma_x(\partial B) - \lambda > 0.$$

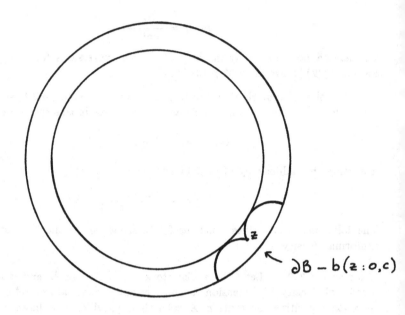

Figure 4.3.2

Now

$$\delta = \sigma_0[b(\gamma(0):0,c)] \ = \ \sigma_0[\gamma(b(0:\gamma^{-1}(0),c))] \ = \ \sigma_{\gamma^{-1}(0)}[b(0:\gamma^{-1}(0),c)] \quad (4.3.2)$$

since σ is a conformal density. However, we also have, from the properties of a conformal density of dimension α, that

$$\sigma_{\gamma^{-1}(0)}[b(0:\gamma^{-1}(0),c)] = \int\limits_{b(0:\gamma^{-1}(0),c)} \left(\frac{d\sigma_{\gamma^{-1}(0)}}{d\sigma_0}\right)(\xi)\, d\sigma_0(\xi)$$

$$= \int\limits_{b(0:\gamma^{-1}(0),c)} P(\gamma^{-1}(0),\xi)^\alpha\, d\sigma_0(\xi). \quad (4.3.3)$$

Our next task is to approximate the Poisson kernel appearing in the above integral.

Lemma 4.3.3   With the above notation there exists a positive constant A depending only on c such that if $\xi \in b(0:\gamma^{-1}(0),c)$ then

$$\frac{A}{1 - |h^{-1}(0)|} < P(\gamma^{-1}(0),\xi) < \frac{2}{1 - |h^{-1}(0)|}.$$

Proof. Note that $P(\gamma^{-1}(0),\xi) = (1 - |h^{-1}(0)|^2)\,|h^{-1}(0) - \xi|^{-2}$ and the upper bound is trivial. To prove the lower bound we note that if $\xi \in b\,(0{:}\gamma^{-1}(0),c)$ then

$$|h^{-1}(0) - \xi| \le 1 - |h^{-1}(0)| + r(0{:}\gamma^{-1}(0),c) \le 1 - |h^{-1}(0)| + \frac{\tanh c\,(1 - |h^{-1}(0)|^2)}{2|h^{-1}(0)|}$$

from (4.3.1). Thus, since $(\gamma^{-1}(0),0) > c$, we see that for a constant B, depending only on c,

$$|h^{-1}(0) - \xi| \le B(1 - |h^{-1}(0)|)$$

which proves the lemma. □

Using this result in equation (4.3.3) we have, for constants a_1, a_2 ,

$$\frac{a_1}{(1 - |h^{-1}(0)|)^\alpha}\sigma_0[b\,(0{:}\gamma^{-1}(0),c)] \le \sigma_{\gamma^{-1}(0)}[b\,(0{:}\gamma^{-1}(0),c)]$$

$$\le \frac{a_2}{(1 - |h^{-1}(0)|)^\alpha}\sigma_0[b\,(0{:}\gamma^{-1}(0),c)].$$

Now we use (4.3.2) to obtain, for positive constants a_3, a_4 ,

$$a_3 < \frac{\sigma_0[b\,(0{:}\gamma^{-1}(0),c)]}{(1 - |h^{-1}(0)|)^\alpha} < a_4.$$

But, for $|h^{-1}(0)|$ close to one, we note from (4.3.1) that

$$(1 - |h^{-1}(0)|)^\alpha \approx r(0{:}\gamma^{-1}(0),c)^\alpha$$

and the proof of the theorem is complete. □

We will be using this theorem to explore the local nature of the measure class of a conformal density. The theorem is clearly saying — at least in a rough sense — that the measure class is behaving somewhat like a Hausdorff α-dimensional measure but only in the neighborhood of limit points which lie in $b\,(0{:}\gamma^{-1}(0),c)$ for **infinitely many** $\gamma \in \Gamma$.

4.4 The Conical Limit Set

In theorem 2.4.4 we showed that if the series

$$\sum_{\gamma \in \Gamma}(1 - |h(a)|)^{n-1}$$

converges then the conical limit set has zero (n-1)-dimensional Lebesgue measure as a subset of ∂B. Our next result is the analogous statement for measures derived from a conformal density.

Theorem 4.4.1   Let Γ be a discrete group acting in B and σ a Γ-invariant conformal density of dimension α. If

$$\sum_{\gamma \in \Gamma} (1 - |\gamma(0)|)^{\alpha} < \infty$$

then the conical limit set has zero σ_x - measure for any $x \in B$.

Proof. This is a standard argument but we include the proof for the sake of completeness. Writing $\Gamma = \{\gamma_n : n = 1,2,...\}$ and choosing $\epsilon > 0$ we find N such that

$$\sum_{n > N} (1 - |\gamma_n(0)|)^{\alpha} < \epsilon.$$

From theorem 4.3.2, using the constant a_4 introduced in the proof of that theorem, we may find N' so that

$$\sum_{n > N'} \sigma_0[b(0{:}\gamma_n(0),c)] < a_4 \epsilon$$

and so

$$\sigma_0 \left\{ \bigcup_{n > N'} [b(0{:}\gamma_n(0),c)] \right\} < a_4 \epsilon.$$

But, from the proof of theorem 2.4.6,

$$C = \bigcup_{c > 0} \bigcap_{N \geq 1} \bigcup_{n > N} b(0{:}\gamma_n(0),c)$$

and so $\sigma_0(C) < a_4 \epsilon$. This is true for every $\epsilon > 0$ and $\sigma_0(C) = 0$ as required. $\square$

Our next result concerns the Hausdorff dimension of the conical limit set: recall from section 1.1 the definition of Hausdorff α-dimensional measure Λ_α.

Theorem 4.4.2   Let Γ be a discrete group acting in B and σ a Γ-invariant conformal density of dimension α. There exists a constant a such that if A is a Borel subset of the conical limit set with $\sigma_x(A) > 0$ then $\Lambda_\alpha(A) \leq a\,\sigma_x(A)$.

Proof. Since $\sigma_x(A) > 0$, almost every point (σ_x) of A is a density point of the measure in the sense that for almost every $a \in A$

$$\lim_{t \to 0} \frac{\sigma_x[B(a,t) \cap A]}{\sigma_x[B(a,t)]} = 1$$

where $B(a,t)$ denotes the ball with Euclidean center a and Euclidean radius t —

see [Federer, 1969 p.158]. Thus, given $\delta > 0$, there is a subset A' of A with $\sigma_z(A - A') < \delta$ and a $t_0 > 0$ such that for all $a \in A'$ and $t < t_0$

$$\frac{\sigma_z[B(a,t)\cap A]}{\sigma_z[B(a,t)]} \geq 1 - \delta. \tag{4.4.1}$$

Now construct a cover of A' by balls $b(0{:}\gamma_i(0),c)$ $i = 1,2,...$ such that $r(0{:}\gamma_i(0),c) > r(0{:}\gamma_{i+1}(0),c)$, $r(0{:}\gamma_1(0),c) < \epsilon$ say, and for each i, the center of $b(0{:}\gamma_{i+1}(0),c)$ is outside the union

$$\bigcup_{k=1}^{i} b(0{:}\gamma_k(0),c).$$

Note that this construction is possible because every conical limit point lies in infinitely many balls $b(0{:}\gamma(0),c)$. The balls with half the radii and the same centers are disjoint. Denote this disjoint union by Ω. Now

$$\sum_i r(0{:}\gamma_i(0),c)^\alpha = 2^\alpha \sum_i [r(0{:}\gamma_i(0),c)/2]^\alpha \tag{4.4.2}$$

and, by theorem 4.3.2, the right hand side above is at most a constant multiple of $\sigma_z(\Omega)$. However, the left side is at least $\Lambda_\alpha^\epsilon(A')$ and so $\Lambda_\alpha^\epsilon(A')$ does not exceed $a\,\sigma_z(\Omega)$. Now observe from (4.4.1) that

$$\sigma_z(\Omega) \leq \frac{\sigma_z(A')}{1-\delta} \leq \frac{\sigma_z(A)-\delta}{1-\delta}.$$

Combining these results we have that $\Lambda_\alpha^\epsilon(A') \leq a\,\dfrac{\sigma_z(A)-\delta}{1-\delta}$. Letting $\epsilon \rightarrow 0$ we obtain $\Lambda_\alpha(A') \leq a\,\sigma_z(A)$, and letting $\delta \rightarrow 0$ we have $\Lambda_\alpha(A) \leq a\,\sigma_z(A)$ as required. $\square$

Corollary 4.4.3   If a Γ-invariant conformal density of dimension α exists and if $d(C)$ denotes the Hausdorff dimension of the conical limit set then $d(C) \leq \alpha$ and, in particular, we always have $d(C) \leq \delta$ where δ is the critical exponent of Γ.

Proof. In the proof above we take $A = C$ and then, without introducing density points, we proceed as before to obtain $\Lambda_\alpha^\epsilon(C) \leq a\,\sigma_z(\Omega)$. But of course, $\sigma_z(\Omega) \leq \sigma_z(\partial B)$ and we have $\Lambda_\alpha(C) \leq a\,\sigma_z(\partial B) < +\infty$. It follows then that $d(C) \leq \alpha$. The last statement of the corollary is immediate since we know there exists a δ-dimensional conformal density. $\square$

We will see later that for large classes of discrete groups we have also the inequality $b\,\sigma_z(A) \leq \Lambda_\alpha(A)$ and this, together with theorem 4.4.2, shows that σ_z really does behave like a Hausdorff measure on the conical limit set.

We now consider an ergodic question.

Theorem 4.4.4   Let Γ be a non-elementary discrete group acting in B and σ a Γ-invariant conformal density of dimension α. If A is a Γ-invariant subset of C then either

- $\sigma_x(A) = 0$ or

- $\sigma_x(A) = \sigma_x(\partial B)$.

Proof. We may as well take $x = 0$ and we remark that σ_0 is not a single atom, otherwise a single atom at ξ implies $\gamma(\xi) = \xi$ for every $\gamma \in \Gamma$ and Γ would be elementary. Suppose that $\sigma_0(A) > 0$ and let ξ be a density point for A. Thus we have a sequence $\{\gamma_n^{-1}(0)\}$ converging to ξ in a cone and

$$\frac{\sigma_0[b\,(0{:}\gamma_n^{-1}(0),c\,)\bigcap A\,]}{\sigma_0[b\,(0{:}\gamma_n^{-1}(0),c\,)]} \to 1. \qquad (4.4.3)$$

Let λ be the largest point mass for σ_0 and, arguing as in the proof of theorem 4.3.2, we see that given $\epsilon > 0$

$$\sigma_0[b\,(\gamma_n\,(0){:}0,c\,)] > \sigma_0(\partial B) - \lambda - \epsilon \qquad (4.4.4)$$

providing c and n are large enough. Note that $b\,(\gamma_n\,(0){:}0,c\,) = \gamma_n\,[b\,(0{:}\gamma_n^{-1}(0),c\,)]$ and so

$$\frac{\sigma_0[b\,(\gamma_n\,(0){:}0,c\,)\bigcap A\,]}{\sigma_0[b\,(\gamma_n\,(0){:}0,c\,)]} = \frac{\sigma_{\gamma_n^{-1}(0)}[b\,(0{:}\gamma_n^{-1}(0),c\,)\bigcap A\,]}{\sigma_{\gamma_n^{-1}(0)}[b\,(0{:}\gamma_n^{-1}(0),c\,)]} \qquad (4.4.5)$$

where we have used the fact that A is Γ-invariant. The right hand side of (4.4.5) may be written

$$\frac{\displaystyle\int\limits_{b\,(0{:}\gamma_n^{-1}(0),c\,)\bigcap A} P\,(\gamma_n^{-1}(0),\eta)^\alpha\,d\,\sigma_0(\eta)}{\displaystyle\int\limits_{b\,(0{:}\gamma_n^{-1}(0),c\,)} P\,(\gamma_n^{-1}(0),\eta)^\alpha\,d\,\sigma_0(\eta)}$$

and this is equal to

$$1 - \frac{\displaystyle\int\limits_{b\,(0{:}\gamma_n^{-1}(0),c\,)\bigcap A'} P\,(\gamma_n^{-1}(0),\eta)^\alpha\,d\,\sigma_0(\eta)}{\displaystyle\int\limits_{b\,(0{:}\gamma_n^{-1}(0),c\,)} P\,(\gamma_n^{-1}(0),\eta)^\alpha\,d\,\sigma_0(\eta)}.$$

We bound this quantity from below using the bounds on the Poisson kernel given in lemma 4.3.3. We see that the right hand side of (4.4.5) is at least

$$1 - \frac{B\sigma_0[b(0{:}\gamma_n^{-1}(0),c)\bigcap A']}{\sigma_0[b(0{:}\gamma_n^{-1}(0),c)]},$$

and in view of (4.4.3) we can bound this below by $1 - \epsilon$ provided n is large enough. Thus for n large enough, using (4.4.5),

$$\sigma_0[b(\gamma_n(0){:}0,c)\bigcap A] \geq (1 - \epsilon)\sigma_0[b(\gamma_n(0){:}0,c)].$$

Combining this with (4.4.4) and the fact that $b(\gamma_n(0){:}0,c)\bigcap A$ is a subset of A we have

$$\sigma_0(A) \geq (1 - \epsilon)[\sigma_0(\partial B) - \lambda - \epsilon]$$

letting $\epsilon \rightarrow 0$ we obtain

$$\sigma_0(A) \geq \sigma_0(\partial B) - \lambda. \qquad (4.4.6)$$

If σ_0 has any atoms then it must have at least two and by (4.4.6) all except one of them would lie in A. However, from theorem 3.5.3, a point mass cannot occur at a conical limit point and so $\lambda = 0$ in (4.4.6). This proves the theorem. □

This result is exactly what we need in order to apply theorem 4.2.1 on the uniqueness of conformal densities of a given dimension. Suppose there is a Γ-invariant conformal density of dimension α - call it σ - and suppose further that $\sigma_z(C) > 0$ then $\sigma_z(C) = \sigma_z(\partial B)$ and Γ is ergodic on ∂B. Thus by theorem 4.2.1 σ is the **unique** conformal density of dimension α. Further, such a σ can have no atoms and the Poincaré series must diverge at the exponent α. Such divergence clearly implies that $\alpha \leq \delta$ where δ is the critical exponent of the group. It turns out that δ is the only dimension for which a conformal density can attach a positive mass to C. In order to prove this we need an estimate on the orbital counting function.

4.5 The Orbital Counting Function

We recall the orbital counting function $N(r,x,y)$ defined in section 1.5 and as an application of the properties of a conformal density we prove the following.

Theorem 4.5.1 Let Γ be a discrete group acting in B and σ a Γ-invariant conformal density of dimension α, then for $x,y \in B$ there exists a constant A depending on Γ, on x, and on y such that, for $r > r_0$ say,

$$N(r,x,y) < Ae^{r\alpha}.$$

Proof. We may as well take $x = y = 0$ and, for integer k, consider the collection $b(0{:}\gamma(0),c)$ for which $k - 1/2 \leq (0,\gamma(0)) < k + 1/2$. If $\xi \in \partial B$ lies in such a $b(0{:}\gamma(0),c)$ then $\gamma(0)$ must be within a hyperbolic distance c of the geodesic segment $\{\lambda\xi : (0,\lambda\xi) \in [k - 1/2, k + 1/2]\}$. This places $\gamma(0)$ in a ball of radius

$1 + c$ and **any** such ball contains at most M images of zero, where M depends only on Γ and c. It follows that the collection $b(0:\gamma(0),c)$ with $k - 1/2 \leq (0,\gamma(0)) < k + 1/2$ covers any point $\xi \in \partial B$ with multiplicity m_ξ and $0 \leq m_\xi \leq M$. Thus

$$\sum_{\gamma \,:\, k-1/2 \,\leq\, (0,\gamma(0)) \,<\, k+1/2} \sigma_0[b(0:\gamma(0),c)] \leq M \, \sigma_0(\partial B). \qquad (4.5.1)$$

Define $n(k)$ to be the cardinality of the set

$$\{\gamma \in \Gamma : k-1/2 \leq (0,\gamma(0)) < k+1/2\}.$$

Using theorem 4.3.2, inequality (4.5.1), and the fact that if $k-1/2 \leq (0,\gamma(0)) < k+1/2$ then, for an absolute constant A, $(1 - |\gamma(0)|)^\alpha$ is greater than $Ae^{-k\alpha}$, we obtain

$$n(k) \, Ae^{-k\alpha} \, a_3 \leq M \, \sigma_0(\partial B)$$

where a_3 is the constant appearing in the proof of theorem 4.3.2. Thus

$$n(k) \leq B \, e^{k\alpha} \qquad (4.5.2)$$

for k large enough, and for a constant B depending on Γ. If we form the sum of terms $n(1), n(2),...,n(R)$ say, we obtain $N(R,0,0)$. Using (4.5.2) this is bounded by a geometric progression whose sum is a constant multiple of $e^{R\alpha}$. $\square$

Corollary 4.5.2   Let Γ be a discrete group acting in B with critical exponent δ. For a constant A depending on Γ, on x, and on y, and for $r \geq r_0$ say,

$$N(r,x,y) \leq A \, e^{r\delta}.$$

Proof. We merely observe that a Γ-invariant conformal density of dimension δ is known to exist. $\square$

Noting that $\delta \leq n-1$ we see that we always do at least as well as the estimate given in theorem 1.5.1 — and better if $\delta < n-1$.

Corollary 4.5.3   Let Γ be a discrete group acting in B, and σ a Γ-invariant conformal density of dimension α, then $\alpha \geq \delta(\Gamma)$.

Proof. The series $\sum_{\gamma \in \Gamma} e^{-s(0,\gamma(0))}$ may be written in terms of an integral as

$$\lim_{R \to \infty} \int_0^R e^{-st} \, dN(t,0,0) = \lim_{R \to \infty} \left[N(R,0,0)e^{-sR} + s \int_0^R e^{-st} N(t,0,0) dt \right].$$

An easy calculation using corollary 4.5.2 shows that this limit is finite provided $s \geq \alpha$. Thus the series $\sum_{\gamma \in \Gamma} e^{-s(0,\gamma(0))}$ converges for $s \geq \alpha$ and it follows that

$\alpha \geq \delta$. □

In the next section we restrict attention to a special class of groups - the convex co-compact groups, and find that we can say much more in this case.

4.6 Convex Co-Compact Groups

The reader will recall the definitions of geometrically finite and convex co-compact discrete groups given in section 1.4. The results of this section were first developed by Sullivan [Sullivan, 1979] for convex co-compact groups and later extended [Sullivan, 1984] to the more general geometrically finite case.

For geometrically finite groups the limit set comprises bounded parabolic fixed points and conical limit points — theorem 2.7.2. From theorem 3.5.10 we note that the measures of M_x have no atomic part and thus assign full measure to the conical limit set. From the results of the preceding sections we see that there is only one Γ-invariant conformal density on a geometrically finite group. Its dimension is δ — the critical exponent of the group — and it has a measure class coinciding with the measures μ_x constructed in chapter 3. Note that these measures are **unique** (up to a multiple) and we shall speak throughout this section of **the** conformal density and **the** measure μ_x.

We will be able to show, in the convex co-compact case, that the measure μ_x is really Hausdorff δ-dimensional measure and that the Hausdorff dimension of the limit set is δ. We further show that the orbital counting function behaves as one would expect — namely like $e^{\delta r}$ to within a bounded multiple. For convex co-compact groups we are able to obtain a great deal of information on the local structure of the measure μ_x. In order to do this we need to show that any ball in ∂B centered at a limit point behaves like a projected ball $b(0{:}\gamma(0),c)$ for some $\gamma \in \Gamma$. The following little lemma is crucial.

Lemma 4.6.1   Let $d > 0$, $c > d$ be given. Then there exist positive constants a_1, a_2, λ such that if $x, y \in B$ with $(x,y) < d$ and $\min \{ (x,0),(y,0) \} \geq \lambda$ then

$$\frac{r(0{:}y,c+d)}{r(0{:}x,c)} < a_1 \quad \text{and} \quad \frac{r(0{:}y,c-d)}{r(0{:}x,c)} > a_2$$

where we are using the notation of section 4.3.

Proof. Using (4.3.1) we see that

$$\frac{\tan r(0{:}y,c+d)}{\tan r(0{:}x,c)} = \frac{\tanh (c+d)}{\tanh c} \frac{1 - |y|^2}{1 - |x|^2} \frac{|x|}{|y|}.$$

We   assume λ is   so   large   that $\tan r(0{:}x,c) < 2r(0{:}x,c)$ when   min

$\{(x,0),(y,0)\} \geq \lambda$. It follows then that

$$\frac{r(0:y,c+d)}{r(0:x,c)} < \frac{\tanh\,(c+d)}{2\tanh c}\,\frac{1-|y|}{1-|x|}$$

Note that

$$\frac{1-|y|}{1-|x|} = \frac{(1+|y|)\,e^{[(0,x)-(0,y)]}}{1+|x|} < 2e^{(x,y)} < 2e^d$$

and with

$$a_1 = e^d\,\frac{\tanh\,(c+d)}{\tanh c}$$

the first part of the lemma is proved. The other half of the proof is entirely similar. □

With the aid of this result we can show that **any** Euclidean ball in ∂B centered at a limit point ξ for Γ has the property of theorem 4.3.2. We denote by $B(\xi,r)$ a Euclidean ball in ∂B which is centered at ξ and has radius r.

Theorem 4.6.2   Let Γ be a convex co-compact group, then there exist constants c,C,r_0 such that if ξ is a limit point for Γ and $r \leq r_0$ then

$$c < \frac{\mu_0[B(\xi,r)]}{r^\delta} < C.$$

Proof. Let D be any fundamental region for $C(\Lambda)$ which contains the origin — actually we cannot assume this, even up to conjugation, but if the origin does not belong to the convex hull of the limit set then we select some fundamental region for $C(\Lambda)$ and consider some domain D' of the form $D' = \{x : \rho(x,C(\Lambda)) < \lambda\}$. If λ is chosen large enough, then the origin belongs to D', and in the following argument D may be replaced by D'. Let d be the non-Euclidean diameter of D. Since ξ is a limit point, the radius from 0 to ξ is covered by Γ-images of D. Fix a value of c ($> d$) so large that $c - d$ satisfies the hypothesis of theorem 4.3.2 for the constant c. Now construct a ball $\Delta\,(k\xi,c)$ centered on the radius to ξ with the property that $b(0:k\xi,c) = B(\xi,r)$. This construction uniquely determines a value of k. Now $k\xi \in \gamma(D)$ for some $\gamma \in \Gamma$ and $(\gamma(0),k\xi) \leq d$ from which it follows that

$$\Delta(\gamma(0),c-d) \subset \Delta(k\xi,c) \subset \Delta(\gamma(0),c+d)$$

and so

$$b(0:\gamma(0),c-d) \subset b(0:k\xi,c) \subset b(0:\gamma(0),c+d)\,.$$

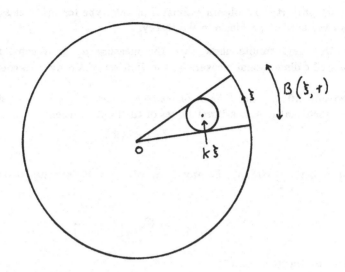

Figure 4.6.1

Thus

$$b(0{:}\gamma(0),c-d) \subset B(\xi,r) \subset b(0{:}\gamma(0),c+d).$$

We obtain the following inequalities

$$\mu_0\left[b(0{:}\gamma(0),c-d)\right] \leq \mu_0\left[B(\xi,r)\right] \leq \mu_0\left[b(0{:}\gamma(0),c+d)\right]. \qquad (4.6.1)$$

Choose $\epsilon > 0$ then for r small enough

$$1 - \epsilon < \frac{r}{r(0{:}k\xi,c)} < 1 + \epsilon.$$

(Recall that $r(0{:}k\xi,c)$ is the radius of $b(0{:}k\xi,c) = B(\xi,r)$ in the d_0 metric - which is locally the Euclidean metric). Thus dividing through (4.6.1) by r^δ we obtain

$$\frac{\mu_0\left[B(\xi,r)\right]}{r^\delta} \leq \frac{\mu_0\left[b(0{:}\gamma(0),c+d)\right]}{r(0{:}\gamma(0),c+d)^\delta} \left[\frac{r(0{:}\gamma(0),c+d)}{r(0{:}k\xi,c)}\right]^\delta \frac{1}{(1-\epsilon)^\delta}$$

with a similar lower bound. By theorem 4.3.2 and lemma 4.6.1 we see that

$$\frac{\mu_0\left[B(\xi,r)\right]}{r^\delta} \leq a_1 A \frac{1}{(1-\epsilon)^\delta}$$

and the proof is complete. $\square$

We have shown that any small ball centered at a limit point has μ_0 measure behaving like r^δ. Our proof of this fact made crucial use of the convex co-

compact property. To obtain estimates of this type for other classes of groups is one of the hardest problems in the theory.

Our next results show that the measure μ_x is a constant multiple of Hausdorff δ-dimensional measure on the limit set of a convex co-compact group.

Theorem 4.6.3   Let Γ be a convex co-compact group, there exists a constant b (> 0) such that if A is a Borel subset of the limit set then

$$b\ \mu_x(A) \leq \Lambda_\delta(A).$$

Proof. Let $\bigcup\limits_{i\,-\,1}^{\infty} B(\xi_i, r_i)$ be any cover of A by balls centered on the limit set. Then

$$\mu_x(A) \leq \sum_{i\,-\,1}^{\infty} \mu_x[B(\xi_i, r_i)]$$

and so, by theorem 4.6.2

$$\mu_x(A) \leq C \sum_{i\,-\,1}^{\infty} r_i^{\delta} \tag{4.6.2}$$

provided the balls all have radius $r_i < r_0$. Since (4.6.2) is true for any such cover of A we see that $\mu_x(A) \leq C\ \Lambda_\delta^\epsilon(A)$ and letting $\epsilon \rightarrow 0$ we obtain the required result. $\square$

If we apply theorem 4.6.3 taking A to be the entire limit set Λ we see that

$$0 < b\ \mu_x(\Lambda) \leq \Lambda_\delta(\Lambda)$$

and so the Hausdorff δ-dimensional measure of Λ is positive. Thus the Hausdorff dimension of the limit set is at least δ. But, from corollary 4.4.3, we note that the Hausdorff dimension of the conical limit set is at most δ in the convex co-compact case. Thus, for convex co-compact groups, the Hausdorff dimension of the limit set is equal to the exponent of convergence.

Theorem 4.6.4   Let Γ be a convex co-compact group. There exists a positive constant k such that

$$\mu_0 = k\ \Lambda_\delta.$$

Proof. For the measure μ_0 we have the transformation rule

$$\mu_0(\gamma(E)) = \int_E |\gamma'(\xi)|^\delta d\mu_x(\xi)$$

where $\gamma \in \Gamma$ and $\delta = \delta(\Gamma)$ is the critical exponent. The Hausdorff δ-dimensional measure also obeys the same transformation rule (in fact for any Moebius γ). We form the new measure

$$M = \mu_0 + \Lambda_\delta$$

and note that it also obeys the same transformation rule. Now μ_0 is absolutely continuous with respect to M and we form the Radon-Nikodym derivative

$$g = \frac{d\mu_0}{dM}$$

on the limit set. The function g is clearly measurable (μ_0) and Γ-invariant. The limit set is made up entirely of conical limit points and Γ is ergodic (μ_0) on the conical limit set (theorem 4.4.4). It follows that g is identically constant. To complete the proof we need only show that this constant is neither 0 nor 1. This means we must show that Λ_δ is neither 0 nor $+\infty$ on the limit set. Theorem 4.4.2 shows that Λ_δ is a finite measure on the limit set and theorem 4.6.3 shows that it is not the zero measure. $\square$

With the convex co-compact hypothesis we can show that the orbital counting function behaves essentially like $e^\delta R$. The reader will recall an old result of Tsuji [Tsuji, 1959 p.518] which states that for a Fuchsian group of finite area,

$$a \; e^r < N(r,0,x) < A \; e^r$$

where A is a group constant and a depends on Γ and on x. If one asks in addition that the group have a compact fundamental region in the unit disc (it is now convex co-compact) then a becomes a group constant. Note that for a Fuchsian group of finite area $\delta = 1$. The following result generalizes Tsuji's estimate.

Theorem 4.6.5   Let Γ be a convex co-compact group with critical exponent δ. There exist positive constants A, r_0 depending only on Γ such that for any $x, y \in B$

$$N(r,x,y) \leq A \; e^{r\delta} \quad \text{for } r > r_0.$$

Further, there exist positive constants a, r_1 depending only on Γ such that for any $x, y \in C(\Lambda)$

$$N(r,x,y) \geq a \; e^{r\delta} \quad \text{for } r > r_1.$$

Proof. Theorem 4.5.1 gives us an upper bound of the required type but with a constant A depending on x and y. The reader will verify from the proof of theorem 4.5.1 that the following lemma yields an A independent of x and y.

Lemma 4.6.6   Let Γ be a convex co-compact group, fix $R > 0$ and for $x \in C(\Lambda)$, $y \in B$ set $M(x,y) = card\{\Gamma(y) \cap \Delta(x,R)\}$ then $M(x,y)$ is uniformly bounded for all such x,y.

Proof. If the lemma is false then there exist sequences $\{x_n\} \subset C(\Lambda)$, $\{y_n\} \subset B$ with $M(x_n,y_n) \to \infty$. Let D be a (compact) fundamental region for $C(\Lambda)$ and, observing that $M(\gamma(x),y) = M(x,y)$ for any $\gamma \in \Gamma$, we may as well assume that $x_n \in D$ for every n. It follows then that $\Delta(x_n,R)$ is, for every n, contained in a fixed compact subset Q of B. It is clear that Q meets only a finite number of Γ-images of D and only a finite number of components of $B - C(\Lambda)$. This contradicts our assertion that

$$card\ \{\Gamma(y_n) \cap Q\} \to \infty \quad \text{as } n \to \infty.$$

Thus we have the upper bound of the theorem. $\square$

With D as in the proof of the lemma let d be its hyperbolic diameter. Select $x,y \in D$, partition the ball B into annuli,

$$U_k = \{z \in B : 3dk \leq (x,z) \leq 3d(k+1)\}$$

and write $n(k)$ for the number of $\gamma \in \Gamma$ with $\gamma(y) \in U_k$. If $\xi \in \Lambda$ then the geodesic joining x to ξ is covered by Γ-images of D. Select x_k on this geodesic with $(x,x_k) = 3d(k+1/2)$ and suppose that $x_k \in \gamma_k(D)$. Now $(x_k,\gamma_k(y)) < d$ and so $\gamma_k(y) \in U_k$. Further, if $c > d$ then $\xi \in b(x{:}\gamma_k(y),c)$. Select c bigger than d and so large that the hypothesis of theorem 4.3.2 is satisfied. Then we have proved that the collection

$$\{b(x{:}\gamma(y),c) : \gamma(y) \in U_k\}$$

covers Λ with multiplicity at least one. This implies

$$\sum_{\gamma\, :\, \gamma(y) \in U_k} \mu_0[b(x{:}\gamma(y),c)] \geq \mu_0(\partial B)$$

and, by theorem 4.3.2 and lemma 4.6.1, there is a group constant A such that

$$A \sum_{\gamma\, :\, \gamma(y) \in U_k} r(x{:}\gamma(y),c)^\delta \geq \mu_0(\partial B). \tag{4.6.3}$$

We now use lemma 4.3.1 to obtain an upper bound for $r(x{:}\gamma(y),c)$ when $\gamma(y) \in U_k$. This is a routine estimate which yields a constant B, depending only on c, for which

$$r(x{:}\gamma(y),c) < Be^{-3dk}.$$

provided $\gamma(y) \in U_k$. Use of this estimate in (4.6.3) yields, for a constant C, $n(k) \geq Ce^{3dk\delta}$ from which, by summing the appropriate geometric series,

$$N(3dk,x,y) \geq C\, e^{3dk\delta}.$$

Thus, for a group constant b,

$$N(r,x,y) \geq b\, e^{r\delta}$$

and this is the required result.$\Box$

4.7 Summary

Let us review the situation to date concerning Γ-invariant conformal densities. To this end suppose Γ is a non-elementary discrete group acting in B and let δ be the critical exponent.

1. If there exists a Γ-invariant conformal density of dimension α then $\alpha \geq \delta$ (corollary 4.5.3).

2. There **does** exist a Γ-invariant conformal density of dimension δ - this is the density associated to the measures μ_x constructed in Chapter 3 (theorem 4.1.2).

3. If $d(C)$ is the Hausdorff dimension of the conical limit set then $d(C) \leq \delta$ (corollary 4.4.3).

4. The action of Γ on C is ergodic with respect to the measure class of any conformal density (theorem 4.4.4).

5. If there exists a Γ-invariant conformal density of dimension α with the property that $\sigma_x(C) > 0$ then

   - $\alpha = \delta$ (corollary 4.5.3 and theorem 4.4.4)

   - σ is the **unique** conformal density of dimension δ - in particular the measure classes M_x constructed in Chapter 3 comprise a **single** measure for each $x \in B$ (theorem 4.2.1)

   - $\sigma_x(C) = \sigma_x(\partial B)$ (theorem 4.4.4)

   - Γ is of divergence type (theorem 4.4.1).

6. If Γ is geometrically finite then the properties of (5) above hold.

7. If Γ is convex co-compact with critical exponent δ then

   - the Hausdorff dimension of the limit set is δ

   - the Hausdorff δ-dimensional measure is, within a bounded multiple, the μ_x measure constructed in chapter 3.

- there is precisely one conformal density invariant by Γ - this is the density μ.

- Γ is of divergence type.

- There exist positive constants a, A, r_0 depending only on Γ such that for $r > r_0$

$$a\ e^{r\delta} \leq N(r,x,y) \leq A\ e^{r\delta}.$$

It is interesting to observe that we have realized δ in two distinct ways

- $\delta = \inf \{s \in R^+ : \sum_{\gamma \in \Gamma} e^{-s(x,\gamma(y))} < \infty\}.$

- $\delta = \inf \{\alpha : \text{there exists a } \Gamma\text{-invariant conformal density of dimension } \alpha\}.$

CHAPTER 5

Hyperbolically Harmonic Functions

5.1 Introduction

We denote by Δ the Laplacian in R^n. A real valued function f defined in B which satisfies the differential equation $\Delta f = 0$ is said to be harmonic. In general we will not be able to use such functions because if γ is a Moebius transform preserving B and f is harmonic in B, the function $f \circ \gamma$ is usually **not** harmonic (except when $n = 2$). We will introduce an operator (the Laplace-Beltrami operator) which leads to a class of functions invariant under composition with Moebius transforms and which enjoys many of the properties of harmonic functions. A full account of the derivation and properties of such functions is to be found in the text of Ahlfors [Ahlfors, 1981] and in the text of Rodin and Sario [Rodin and Sario, 1968]. In this section we summarize the results we need. For the most part these are taken verbatim from Ahlfors' text to which the reader is referred for the proofs.

Definition The Laplace-Beltrami operator Δ_2 for the unit ball B of R^n equipped with the metric ρ is given by

$$\Delta_2 = \frac{(1-r^2)^2}{4} \left[\Delta + \frac{2(n-2)r}{1-r^2} \frac{\partial}{\partial r} \right]$$

where $r = |x|$. The upper half-space version of this operator (also denoted Δ_2) is given by

$$\Delta_2 = x_n^2 \left[\Delta - \frac{n-2}{x_n} \frac{\partial}{\partial x_n} \right]$$

Definition A function f in B (or in H) which satisfies $\Delta_2 f = 0$ is called hyperbolically harmonic — henceforth abbreviated to h.h.

This is the class of functions we are after. The crucial fact that this class is invariant under composition with Moebius functions is proved in [Ahlfors, 1981 p.55]. In fact even more than this is true.

Theorem 5.1.1   For a function f in B (or in H), if γ is a Moebius transform preserving B (or H) then $\Delta_2(f \circ \gamma) = (\Delta_2 f)\circ\gamma$.

We note from the definition of Δ_2, and the theorem above, that harmonic functions remain harmonic when composed with a Moebius transform — provided $n = 2$.

We next give some examples of h.h functions and consider first those which depend only on r. A fairly straightforward calculation [Ahlfors, 1981 p.57] proves the following.

Theorem 5.1.2   A function u depending only on r and h.h in $0 < \epsilon < r < 1$ is of the form

$$u(r) = a \int_\epsilon^r \frac{(1 - t^2)^{n-2}}{t^{n-1}} \, dt + b$$

where a and b are real constants.

We note that no such function can stay finite at $r = 0$. We may write the normalized solution

$$g(r) = \int_r^1 \frac{(1 - t^2)^{n-2}}{t^{n-1}} \, dt \qquad (5.1.1)$$

and so for $n = 2$, $g(r) = -\log r$, and for $n = 3$, $g(r) = \frac{1}{r} + r - 2$. In general we have the estimates (for $n > 2$) $g(r) \sim \frac{r^{2-n}}{n-2}$ as $r \to 0$ and and $g(r) = O((1-r)^{n-1})$ as $r \to 1$. A further example of an h.h function is given by the following result.

Theorem 5.1.3   Suppose $|\xi| = 1$ and $|x| < 1$ and set

$$P(x,\xi) = \frac{1 - |x|^2}{|x - \xi|^2}$$

then $(P(x,\xi))^\alpha$ is an eigenfunction for Δ_2 with eigenvalue $\alpha(\alpha - n + 1)$. In the upper half-space model, the corresponding eigenfunction is x_n^α.

Proof. Considering the upper half-space situation

$$\Delta_2(x_n^\alpha) = x_n^2 \left[\Delta(x_n^\alpha) - \frac{n-2}{x_n} \frac{\partial}{\partial x_n} (x_n^\alpha) \right]$$

$$= x_n^2 \left[\alpha(\alpha-1)x_n^{\alpha-2} - \frac{(n-2)}{x_n} \alpha x_n^{\alpha-1} \right]$$

$$= \alpha(\alpha-1) x_n^\alpha - \alpha(n-2)x_n^\alpha = \alpha(\alpha-n+1) x_n^\alpha$$

as required. Now recall the map V from B to H defined in section 1.3. Note that for $y = (y_1, y_2, ..., y_n)$ in B

$$\left(V(y) \right)_n = \frac{1 - |y|^2}{|y - e_n|^2}$$

and so $P(y, e_n)^\alpha$ is an eigenfunction for Δ_2 with eigenvalue $\alpha(\alpha - n + 1)$. Now given any $x \in \partial B$ there exists a rotation β with $\beta x = e_n$. However, it is obvious that

$$P(y, x) = P(y, \beta^{-1}(e_n)) = P(\beta y, e_n) .$$

By theorem 5.1.1 we conclude that $\Delta_2 P(y, x)^\alpha = \alpha(\alpha - n + 1) P(y, x)^\alpha$, and the theorem is proved. □

The following result [Ahlfors, 1981 p.69] is exactly what one would expect.

Theorem 5.1.4   Suppose f is an L_1 function defined on S then the function

$$u(x) = \frac{1}{w(S)} \int_S (P(x, \xi))^{n-1} f(\xi) dw(\xi)$$

is h.h in B.

The function u defined in this theorem has the expected property with respect to boundary approach [Ahlfors, 1981 p.69].

Theorem 5.1.5   Let f, u be given as in theorem 5.1.4 and suppose $\xi \in S$ then, for almost all (w) such ξ, $u(x)$ converges to $f(\xi)$ as x approaches ξ in a cone $|x - \xi| < M(1 - |x|)$.

As an immediate consequence of (1.3.2) and theorem 1.3.3 we have that

$$P(x, \xi) = P(\gamma(x), \gamma(\xi)) |\gamma'(\xi)| \tag{5.1.2}$$

for any Moebius γ preserving B. Now if f and u are as in theorem 5.1.4 then, using (5.1.2) we obtain

$$u(\gamma(x)) \;=\; \frac{1}{w(S)} \int_S (P(x,\xi))^{n-1} f(\gamma(\xi)) \, dw(\xi) \tag{5.1.3}$$

and we note in particular that if $f(\gamma(\xi)) = f(\xi)$ for almost every (w) $\xi \in S$ then $u(\gamma(x)) = u(x)$. It is an immediate observation that if the function f is constant almost everywhere then u is identically equal to this constant.

The Fatou theorem may be proved using the representation given in theorem 5.1.4 exactly as is done for harmonic functions.

Theorem 5.1.6   Let u be a bounded h.h function in B. For almost every $\xi \in S$, $\displaystyle \lim_{t \to 1} u(t\xi) = f(\xi)$ exists and

$$u(x) \;=\; \frac{1}{w(S)} \int_S (P(x,\xi))^{n-1} f(\xi) \, dw(\xi).$$

We also have the h.h version of Green's formula. In order to state this it will be convenient to replace all terms by their hyperbolic counterparts.

- volume element $dV = \dfrac{2^n \, dx_1 \cdots dx_n}{(1 - |x|^2)^n}$

- area element $d\sigma_h = \dfrac{2^{n-1} d\sigma}{(1 - |x|^2)^{n-1}}$

- normal derivative $\dfrac{\partial v}{\partial n_h} = \dfrac{1 - |x|^2}{2} \, \dfrac{\partial v}{\partial n}$

- gradient $\nabla_h u = \dfrac{1 - |x|^2}{2} \, \nabla u$

- Laplacian $\Delta_h v = \Delta_2 v$

With these notations we have [Ahlfors, 1981 p.62].

Theorem 5.1.7   Let u, v be real valued functions in B with continuous second partials and $D \subset\!\subset B$ with smooth boundary then

$$\int_D u \, \Delta_h v \, dV \;=\; \int_{\partial D} u \, \frac{\partial v}{\partial n_h} \, d\sigma_h \;-\; \int_D (\nabla_h u \cdot \nabla_h v) \, dV_h.$$

The more common formula is as follows.

Theorem 5.1.8 With u,v and D as in theorem 5.1.7

$$\int_D (u\,\Delta_h v \;-\; v\,\Delta_h u)\,dV \;=\; \int_{\partial D} (u\,\frac{\partial v}{\partial n_h} \;-\; v\,\frac{\partial u}{\partial n_h})\,d\sigma_h.$$

We next wish to consider the Green's function. Let Γ be a discrete group acting in B and we write $M(\Gamma)$ for the Hausdorff topological space B/Γ. A function on $M(\Gamma)$ can be viewed as the projection of a Γ-invariant function on B. Following Ahlfors [Ahlfors, 1981 p.88] we define the Green's function on $M(\Gamma)$ as follows.

Definition Let Γ be a discrete group acting in the unit ball B and suppose $a \in B$. If there is a function G from $B-\Gamma(a)$ to R such that

- G is h.h on $B-\Gamma(a)$

- $G \circ \gamma \equiv G$ for any $\gamma \in \Gamma$

- G has the singularity given by (5.1.1) at a

- G is the smallest positive function with these properties
then the Green's function on $M(\Gamma)$ with pole at the projection of a is defined to be the projection of G.

The following result is proved in [Ahlfors, 1981 p.88].

Theorem 5.1.9   Let Γ be a discrete group acting in the unit ball B then $M(\Gamma)$ has a Green's function if and only if Γ converges at the exponent $n-1$.

It is customary, using the terminology of Riemann Surface theory, to say that the quotient space $M(\Gamma)$ belongs to the class O_G if it does not support a Green's function. Thus the previous theorem states that $M(\Gamma) \in O_G$ if and only if Γ diverges at the exponent $n-1$.

5.2 Harmonic Measure

We have seen that the Laplace-Beltrami operator is defined on $M(\Gamma)$ and in fact we can solve the Dirichlet problem for regular regions and continuous boundary functions [Rodin and Sario, 1968 p.238].

Let $\{M_n\}_0^\infty$ be an exhaustion of $M(\Gamma)$ with connected $M-\overline{M_0}$. For each n we let $w_n(x)$ be the continuous function on $M(\Gamma)$ with $w_n\,|\,\overline{M_0} = 0$, $w_n\,|\,M(\Gamma) - M_n = 1$, and w_n h.h in $M_n - \overline{M_0}$. By the maximum principle, $w_{n+p} \leq w_n$ and thus $\lim_{n \to \infty} w_n(x)$ exists on $M(\Gamma)$, vanishes on $\overline{M_0}$ and is h.h on $M(\Gamma) - \overline{M_0}$. This function is denoted $w(x)$ and is referred to as the harmonic measure of the ideal boundary of $M(\Gamma)$ with respect to $M(\Gamma) - \overline{M_0}$.

It is a well known result for Riemann Surfaces that the harmonic measure of the ideal boundary is identically zero if and only if the surface does not have a Green's function. The proof of this fact [Ahlfors and Sario, 1960 p.204] generalizes easily and we have the following.

Theorem 5.2.1   Let Γ be a discrete group acting in the unit ball B then the following are equivalent

- Γ diverges at the exponent $n-1$

- $M(\Gamma)$ has no Green's function

- The harmonic measure of the ideal boundary of $M(\Gamma)$ is identically zero.

Another classical theorem for Riemann surfaces, due to P.J. Myrberg, states that if a surface has a non-trivial bounded harmonic function then it has a Green's function and again the proof generalizes easily [Ahlfors and Sario, 1960 p.204].

Theorem 5.2.2   Let Γ be a discrete group acting in the unit ball B, if $M(\Gamma)$ has no Green's function then any bounded h.h function on $M(\Gamma)$ reduces to a constant.

The next result generalizes a theorem of Seidel [Seidel, 1935].

Theorem 5.2.3   Let Γ be a discrete group acting in the unit ball B. There exists a measurable, Γ-invariant subset A of S with $w(A) > 0$ and $w(\tilde{A}) > 0$ if and only if $M(\Gamma)$ has a bounded non-trivial h.h function.

Proof. Suppose first that a set A with the properties stated in the theorem exists and set 1_A to be the characteristic function of A defined on S. We define

$$u(x) = \frac{1}{w(S)} \int_S (P(x,\xi))^{n-1} 1_A(\xi) \, dw(\xi)$$

which, by theorem 5.1.4 is h.h in B and, by theorem 5.1.5, is non-constant in B. Next observe from (5.1.3) that u is Γ-invariant, and so we have a non trivial bounded h.h function on $M(\Gamma)$. The reverse conclusion follows just as easily from the theorems in section 5.1.

Corollary 5.2.4   Let Γ be a discrete group acting in the unit ball B. If Γ diverges at the exponent $n-1$ and if A is a measurable Γ-invariant subset of S then either $w(A) = 0$ or $w(\tilde{A}) = 0$.

In the terminology of the next chapter, groups which diverge at the exponent $n-1$ act **ergodically** on the sphere at infinity.

We conclude this section with a test for determining whether a group Γ diverges at the exponent $n-1$. This test generalizes the Euclidean metric test of Laasonen for Riemann surfaces with no Green's function [Sario and Nakai, 1970 p.330].

Theorem 5.2.5   Let Γ be a discrete group acting in the unit ball B and let D_0 be the Dirichlet region for Γ centered at the origin. Set $\theta(r) = D_0 \cap \{x: |x| = r\}$ for $0 < r < 1$. If

$$\int_\epsilon^1 \frac{(1-r)^{n-2}}{w(\theta(r))} \, dr = \infty$$

then Γ diverges at the exponent $n-1$.

Proof. Suppose that Γ converges at the exponent $n-1$ then, from the results above, the harmonic measure of the ideal boundary of $M(\Gamma)$ is non-zero. To be precise, if $\Delta(x,\delta)$ denotes a closed hyperbolic ball centered at $x \in B$ of radius δ then there exists a function h, defined in B, with the following properties :

1. h is continuous and bounded in B

2. h is invariant under Γ

3. h is h.h on $B - \bigcup_{\gamma \in \Gamma} \Delta(\gamma(0),\delta)$

4. if $x \in \Delta(\gamma(0),\delta)$, for some $\gamma \in \Gamma$, then $h(x) = 0$.

For $0 < R < 1$ we write $D(R) = D \cap \{x: |x| < R\}$ and $D^*(R)$ is the part of $D(R)$ which lies in $\{x: \rho(x,0) > \delta\}$. Now apply Green's formula to the function $v = h$ in the domain $D^*(R)$ to obtain

$$\Theta(R) = \int_{D^*(R)} \frac{\nabla^2 h}{(1-|x|^2)^{n-2}} \, dV_e = \int_{\partial D^*(R)} h \, \frac{\partial h}{\partial n} \, \frac{dA}{(1-|x|^2)^{n-2}} \quad (5.2.1)$$

where dA and dV_e denote respectively the elements of Euclidean area and volume at the point $x \in B$. Note that h is zero on $\partial\Delta(0,\delta)$ and that h has equal and $\partial h / \partial n$ has opposite values at equivalent boundary points. Therefore we have

$$\Theta(R) = \int_{\theta(R)} h \frac{\partial h}{\partial n} \frac{dA}{(1-R^2)^{n-2}}. \quad (5.2.2)$$

Using polar coordinates (see section 1.1) we obtain from (5.2.1)

$$\Theta(R) = \int_\epsilon^R \int_{\theta(R)} \frac{\nabla^2 h}{(1-r^2)^{n-2}} r^{n-1} (\sin\theta_1)^{n-2} \cdots \sin\theta_{n-2} \, d\theta_1 d\theta_2 \cdots d\theta_{n-1} dr$$

and so

$$\Theta'(R) = \int_{\theta(R)} \frac{\nabla^2 h}{(1-R^2)^{n-2}} R^{n-1} \, dw = \int_{\theta(R)} \frac{\nabla^2 h}{(1-R^2)^{n-2}} \, dA.$$

Thus

$$\Theta'(R) \geq \int_{\theta(R)} \left(\frac{\partial h}{\partial r}\right)^2 \frac{dA}{(1-R^2)^{n-2}}. \qquad (5.2.3)$$

Using the Schwarz inequality we obtain from (5.2.2)

$$\Theta(R)^2 \leq (1-R^2)^{-(n-2)} \int_{\theta(R)} h^2 \, dA \int_{\theta(R)} \left(\frac{\partial h}{\partial r}\right)^2 \frac{dA}{(1-R^2)^{n-2}}.$$

and, from (5.2.3),

$$(1-R^2)^{n-2} \Theta(R)^2 \leq \Theta'(R) \int_{\theta(R)} h^2 \, dA.$$

Since h is bounded in B we have, for some constant M,

$$(1-R^2)^{n-2} \Theta(R)^2 \leq M \, \Theta'(R) \, w(\theta(R))$$

and we see that

$$\int_{\epsilon}^{R} \frac{(1-r^2)^{n-2}}{w(\theta(r))} \, dr \leq M \int_{\epsilon}^{R} \frac{\Theta'(R)}{\Theta(r)^2} \, dr = M \left[\frac{1}{\Theta(\epsilon)} - \frac{1}{\Theta(R)}\right]$$

and the right hand side is bounded by $\dfrac{M}{\Theta(\epsilon)}$ as $R \to 1$. Thus if Γ converges at the exponent $n-1$ the integral

$$\int_{\epsilon}^{R} \frac{(1-r)^{n-2}}{w(\theta(r))} \, dr$$

remains bounded as $R \to 1$ and the theorem is proved . $\square$

We remark that this result has been used [Nicholls, 1981b] to establish the existence of non geometrically finite Kleinian groups which diverge at the exponent 2.

5.3 Eigenfunctions

In this section we consider briefly eigenfunctions of the operator Δ_2. The spectrum of this operator is currently the subject of intense investigation. We shall do little more than touch on this important theory — the reader is referred to Patterson [Patterson, 1987] for an account of the development of the theory and for several further references.

Our starting point is theorem 5.1.3 in which we proved that $(P(x,\xi))^\alpha$ is an eigenfunction for Δ_2 with eigenvalue $\alpha(\alpha-n+1)$. Now suppose that Γ is a discrete group preserving B and that σ is a conformal density for Γ of dimension α. If f is an $L_1(\sigma_0)$ function defined on S then we note that the function u defined in B by

$$u(x) = \int_S (P(x,\xi))^\alpha f(\xi)\, d\sigma_0(\xi) \tag{5.3.1}$$

satisfies $\Delta_2 u = \alpha(\alpha - n + 1)\, u$. Note that

$$u(x) = \int_S (P(x,\xi))^\alpha f(\xi)\, d\sigma_0(\xi) = \int_S (P(x,\xi)/P(0,\xi))^\alpha f(\xi)\, d\sigma_0(\xi)$$

$$= \int_S f(\xi)\, d\sigma_x(\xi)$$

from the properties of a conformal density. It is now immediate, again from the properties of a conformal density, that for $\gamma \in \Gamma$

$$u(\gamma(x)) = \int_S f(\xi)\, d\sigma_{\gamma(x)}(\xi) = \int_S f(\gamma(\xi))\, d\sigma_x(\xi).$$

In other words, the function u is invariant under Γ provided f is. Summarizing these results we have the following result.

Theorem 5.3.1 Let Γ be a discrete group preserving B and σ a Γ-invariant conformal density of dimension α. If f is an $L_1(\sigma_0)$ function on S which is invariant under Γ and if u is defined by (5.3.1) then

- u is an eigenfunction of Δ_2 with eigenvalue $\alpha(\alpha - n + 1)$.

- u is invariant under Γ.

Now we specialize the definition (5.3.1) by taking $f \equiv 1$ and by considering the conformal density μ of dimension $\delta(\Gamma)$. We write

$$\phi_\mu(x) = \int_S P(x,\xi)^\delta\, d\mu_0(\xi) \tag{5.3.2}$$

and note that $\phi_\mu(x) = \mu_x(S)$. The following little estimate will be very useful to us later on.

Lemma 5.3.2 With ϕ_μ defined as above, and with $s_{\xi\eta}(x)$ denoting the hyperbolic distance from the point x to the geodesic connecting ξ and η

$$\left[\phi_\mu(x)\right]^2 = 4^\delta \int_{S\times S} |\xi - \eta|^{-2\delta} [\cosh s_{\xi\eta}(x)]^{-2\delta}\, d\mu_0(\xi)\, d\mu_0(\eta).$$

Proof. From the definition,

$$[\phi_\mu(x)]^2 = \int_S \left(\frac{|1 - |x|^2}{|x - \xi|^2}\right)^\delta d\mu_0(\xi) \cdot \int_S \left(\frac{1 - |x|^2}{|x - \eta|^2}\right)^\delta d\mu_0(\eta)$$

$$= \int_{S \times S} \frac{(1 - |x|^2)^{2\delta}}{|x - \xi|^{2\delta}|x - \eta|^{2\delta}} d\mu_0(\xi)d\mu_0(\eta).$$

Using theorem 1.2.1,

$$\left[\frac{(1 - |x|^2)}{|x - \xi||x - \eta|}\right]^{2\delta} = \frac{4^\delta}{|\xi - \eta|^{2\delta}[\cosh s_{\xi\eta}(x)]^{2\delta}}$$

and, from the above,

$$[\phi_\mu(x)]^2 = 4^\delta \int_{S \times S} \frac{1}{|\xi - \eta|^{2\delta}[\cosh s_{\xi\eta}(x)]^{2\delta}} \, d\mu_0(\xi)d\mu_0(\eta).$$

This is the required result. $\square$

We conclude this section with the important result that, for a geometrically finite group, the function ϕ_μ is square integrable. Let Γ be a geometrically finite group acting on B. The Dirichlet region for Γ centered at the origin will be denoted D_0 and we intersect D_0 with the set of points which are at most a hyperbolic distance of K from the convex hull of the limit set of Γ — this intersection will be written $D(K)$. We need some more notation. For $\xi, \eta \in S$ and $K > 0$ write $D_{\xi,\eta}(K)$ to be the intersection of $D(0)$ with the set of points distant at most K from the geodesic joining ξ and η. It is clear that

$$D(K) = \bigcup D_{\xi,\eta}(K)$$

where the union is taken over those pairs (ξ, η) with the property that the geodesic joining ξ to η lies in the convex hull of the limit set.

Theorem 5.3.3   If Γ is a non-elementary geometrically finite group then, with $D(K)$ defined as above,

$$\int_{D(K)} \left[\phi_\mu(x)\right]^2 dV < \infty.$$

Proof. As a first step, we obtain an estimate on the size of $\phi_\mu(x)$ in a neighborhood of a parabolic cusp. We may as well suppose that our group acts in the upper half-space of R^n and that ∞ is a parabolic cusp of rank k. Since the group is assumed to be geometrically finite we know that the canonical measures μ_x have no atomic part — theorem 3.5.10. However, the limit set comprises conical limit points and parabolic fixed points — theorem 2.7.2 — and we deduce that the conical limit set has full measure. By theorem 4.4.1 the group is of divergence type and so, from the proof of theorem 3.5.9, if $\epsilon > 0$ is chosen,

$$\mu_{x_z}(\overline{H}) < Cz^{k+\epsilon-\delta}.$$

where H denotes the upper half-space and x_z denotes the point $(0,0,...,0,z)$. Thus, with ∞ a parabolic cusp of rank k we have

$$\phi_\mu(x_z) \leq Cz^{k+\epsilon-\delta}. \tag{5.3.3}$$

Let us now estimate the integral

$$\int [\phi_\mu(x)]^2 dV$$

over the piece of $D(K)$ near to a parabolic cusp of rank k. We assume the cusp is at ∞. Recall from lemma 3.5.4 that $\delta > k/2$, and we choose $\epsilon = 1/2(\delta-k/2)$. With this choice of ϵ in (5.3.3) and using theorem 2.7.3 we find that the integral above is bounded by

$$A \int_M^\infty z^{-(1+\delta-k/2)} dz$$

which is finite. Our integral is thus bounded in a neighborhood of any parabolic cusp and, obviously, bounded in any compact piece of $D(K)$. This completes the proof of the theorem. □

CHAPTER 6

The Sphere at Infinity

6.1 Introduction

Suppose Ω is a complete separable metric space with a measure defined on its Borel subsets. Suppose further there is a group Γ of maps of Ω onto itself which are invertible, measurable (if $A \subset \Omega$ is measurable and $\gamma \in \Gamma$ then $\gamma(A)$ is measurable), and non-singular (if $A \subset \Omega$ is of measure zero and $\gamma \in \Gamma$ then $\gamma(A)$ is of measure zero). Typically we will take Ω to be S or $S \times S$ or $S \times S \times S$ with metric derived from the chordal metric on S and measure derived from w measure on S. The group Γ of self maps will typically be a discrete group of Moebius transforms. In this section we gather some definitions and preliminary results.

For $x \in \Omega$ the set $\{\gamma(x): \gamma \in \Gamma\}$ is the **trajectory** of x. If $A \subset \Omega$ has the property that $\gamma(A) = A$ for every $\gamma \in \Gamma$ then A is **invariant** under Γ. A **wandering set** W for Γ on Ω is a measurable subset of Ω with the property that for any $\gamma \in \Gamma - I$ the intersection $\gamma(W) \cap W$ is of measure zero. If there exists a wandering set W of positive measure then the action of Γ is **dissipative**, otherwise the action is **conservative**. The action is said to be **completely dissipative** if there exists a wandering set W of positive measure such that, defining W^* by

$$W^* = \bigcup_{\gamma \in \Gamma} \gamma(W),$$

then $\Omega - W^*$ is of measure zero.

The group Γ is **regionally transitive** on Ω if there exists a point $x \in \Omega$ whose trajectory is dense in Ω. The group is **topometric transitive** on Ω if the set of points which do not have a dense trajectory is of zero measure in Ω. The group is **ergodic** on Ω if, whenever A is a measurable invariant subset of Ω, either A or its complement is of measure zero.

Theorem 6.1.1   The following are equivalent:

1. Γ is regionally transitive on Ω.

2. Every open set in Ω which is invariant under Γ is everywhere dense in Ω.

3. If A, A^* are open sets in Ω then, for some $\gamma \in \Gamma$, $\gamma(A) \bigcap A^* \neq \varnothing$

Proof. To prove that (1) implies (3) we let x be a point with dense trajectory and note that if A, A^* are open, then for some $\gamma_1, \gamma_2 \in \Gamma$ we will have $\gamma_1(x) \in A$, $\gamma_2(x) \in A^*$. It follows that $\gamma_2 \gamma_1^{-1}(A) \bigcap A^* \neq \varnothing$ as required. The fact that (3) implies (2) is trivial, so we only have to show that (2) implies (1). To prove this let $U_1, U_2, ...$ be a base for the open sets in Ω and note that a point x fails to have a dense trajectory if and only if x lies in some set A_m where

$$A_m = \bigcap_{\gamma \in \Gamma} \gamma(\Omega - U_m).$$

The complement of A_m is the set $\bigcup_{\gamma \in \Gamma} \gamma(U_m)$ which is open and Γ-invariant and so, by part (2), is everywhere dense in Ω. It follows that A_m is nowhere dense. If there is no point with a dense trajectory then Ω is a countable union of nowhere dense subsets — which contradicts the Baire category theorem. $\square$

Theorem 6.1.2   The following are equivalent:

1. Γ is topometric transitive on Ω.

2. Every measurable subset of Ω which is of positive measure and is invariant under Γ is everywhere dense in Ω.

3. If M is a measurable subset of Ω which is of positive measure and if D is an open subset of Ω then, for some $\gamma \in \Gamma$, $\gamma(M) \bigcap D \neq \varnothing$.

Proof. To prove that (1) implies (3) we suppose that M is as given in (3) and select $x \in M$ with dense trajectory. If D is any open set then, for some $\gamma \in \Gamma$, $\gamma(x) \in D$ and (3) is proved. The fact that (3) implies (2) is trivial. So we have only to prove that (2) implies (1). With A_m defined as in the proof of theorem 6.1.1 we note that A_m is measurable and Γ-invariant. If, for some m, A_m is of positive measure then from (2) we see that $A_m \bigcap U_m \neq \varnothing$ — but this contradicts the definition of A_m. Thus each A_m is of zero measure and we see that the set of points which fail to have a dense trajectory is a set of measure zero. Thus (1) is true and the proof of the theorem is complete. $\square$

Theorem 6.1.3   The following are equivalent.

1. Γ is ergodic on Ω.

2. If M, M^* are two measurable subsets of Ω each of positive measure then, for some $\gamma \in \Gamma$, $\gamma(M) \cap M^*$ is of positive measure.

Proof. This result follows easily from the fact that a countable union of measurable sets is measurable. $\square$

We remark that, as a consequence of these results, a group which is ergodic on Ω must be both topometric transitive and conservative. Also, a group which is topometric transitive must be regionally transitive.

For the remainder of this chapter we will be considering the action of discrete Moebius groups on the unit sphere and products of the sphere with itself. This action is defined as follows. For γ a Moebius transform preserving B we define

$$\gamma(\xi_1, \xi_2, \ldots, \xi_k) = (\gamma(\xi_1), \gamma(\xi_2), \ldots, \gamma(\xi_k)) \qquad |\xi_i| = 1, \quad i = 1, 2, \ldots, k.$$

This gives us the action of a Moebius transform on the k-fold product $S^{(k)} = S \times S \times S \cdots \times S$ (k factors).

Theorem 6.1.4   If a discrete group Γ is either regionally transitive, topometric transitive, ergodic, or conservative in its action on $S^{(k+1)}$ then it possesses that same property in its action on $S^{(k)}$.

Proof. To prove the theorem as regards regional and topometric transitivity we merely remark that if $(\xi_1, \xi_2, \ldots, \xi_{k+1})$ has a dense trajectory in $S^{(k+1)}$ then $(\xi_1, \xi_2, \ldots, \xi_k)$ clearly has a dense trajectory in $S^{(k)}$. Now suppose that Γ is not ergodic on $S^{(k)}$ and let A be a measurable set of positive measure invariant under Γ and whose complement has positive measure. Considering the set $A \times S$ we see that Γ is not ergodic on $S^{(k+1)}$. Similarly, a wandering set on $S^{(k)}$ gives rise to a wandering set on $S^{(k+1)}$ and the proof of the theorem is complete. $\square$

6.2 Action on S

Our first result is very elementary and characterizes regional and topometric transitivity.

Theorem 6.2.1   The following are equivalent:

1. Γ is topometric transitive on S.

2. Γ is regionally transitive on S.

3. Γ is of the first kind.

Proof.   The fact that (1) implies (2) is immediate. To prove that (2) implies (3) we note that if the trajectory of some point z is dense in S then each point of S is a limit point for Γ which is therefore a group of the first kind. To show that (3) implies (1) we note that for a group of the first kind the limit set is S and that for each $z \in S$ the trajectory is dense in the limit set. $\square$

Our next result (due to Seidel [Seidel, 1935] in dimension 2) characterizes groups which are ergodic on S — this is just theorem 5.2.3.

Theorem 6.2.2   The group Γ is ergodic on S if and only if any bounded Γ-invariant h.h function in B reduces to a constant.

In view of its importance for our later work, we restate corollary 5.2.4.

Theorem 6.2.3   If Γ is a discrete group which diverges at the exponent $n-1$ then Γ is ergodic on S.

We now consider the conservative and dissipative action of Γ on the sphere. The following two theorems are due to Sullivan [Sullivan, 1981]. We recall that H denotes the set of horospherical limit points and it is invariant under Γ. Clearly H is a measurable subset of S, and in fact Γ is conservative in its action on H.

Theorem 6.2.4   The action of Γ is conservative on H.

Proof.   Suppose that H contains a wandering set A. It follows that

$$\sum_{\gamma \in \Gamma} \int_A |\gamma'(z)|^{n-1}\, dw \;<\; \infty$$

and so

$$\int_A \sum_{\gamma \in \Gamma} |\gamma'(z)|^{n-1}\, dw \;<\; \infty.$$

Thus $\sum_{\gamma \in \Gamma} |\gamma'(z)|^{n-1}$ converges almost everywhere in A. However, from theorem 2.5.3 we note that this sum diverges for all $z \in H$. It follows that A has measure zero and the proof is complete. $\square$

In its action on $\tilde{H}$, Γ is dissipative. In fact more than this is true. We can find a measurable subset of $\tilde{H}$ which contains no two Γ-equivalent points and with the further property that the union of its Γ-images comprises all of $\tilde{H}$ with the exception of a set of measure zero. Such a set is called a **fundamental region** for $\tilde{H}$.

Theorem 6.2.5   There is a fundamental region for the group action on $\tilde{H}$.

Proof. Choose $a \in B$ not fixed by any element of Γ and, writing D_a for the Dirichlet region centered at a, we set $e_a = \partial D_a \cap S$. As in the proof of theorem 2.6.4 we see that for any $\gamma \in \Gamma$ the set $e_a \cap \gamma(e_a)$ is countable. Thus we may find a subset e_a^* of e_a which contains no two Γ-equivalent points and which differs from e_a by a countable set. The set $\bigcup_{\gamma \in \Gamma} \gamma(e_a^*)$ covers the Dirichlet set D (except possibly for a set of measure zero) and thus also covers $\tilde{H}$ except for a set of measure zero (theorem 2.6.6). We have shown that e_a^* is a fundamental region for the action of Γ on $\tilde{H}$ as required. $\square$

The construction given in the last theorem can be realized geometrically as follows. Given $a \in B$ the set e_a^* has the following property. The point $\xi \in S$ lies in e_a^* if and only if the closed horoball at ξ through a meets the orbit of a precisely in a. Taking a to be the origin (which we assume is not a fixed point) we note that such a horoball meets the orbit of 0 precisely in 0 if and only if all the group images of the bounding horosphere (except under the identity transform) have strictly smaller radius. We recall the formula for the radius of an image horosphere given after theorem 2.6.3 and we have the following result.

Theorem 6.2.6   For a discrete group Γ the set

$$\{\xi \in S: |\gamma'(\xi)| < 1 \text{ all } \gamma \in \Gamma - I\}$$

is a fundamental region for the action of Γ on $\tilde{H}$ — provided the origin is not an elliptic fixed point.

If the origin is an elliptic fixed point then, for any γ in the stabilizer of 0 and any $\xi \in S$, $|\gamma'(\xi)| = 1$ (from lemma 1.3.1), and so the set given in theorem 6.2.6 is empty. The way to achieve a fundamental region in this case is to start with the set $\{\xi \in S : |\gamma'(\xi)| < 1 \text{ all } \gamma \in \Gamma - stab(0)\}$ and intersect this with a fundamental domain for the action of the stabilizer of 0 on S.

We generalize a definition of Pommerenke [Pommerenke, 1976] to n dimensions and say that a group Γ is of **accessible type** if for some $\beta > 0$ there exists a measurable set $A \subset S$ containing no two Γ-equivalent points such that

$$\sum_{\gamma \in \Gamma} w(\gamma(A)) = \beta.$$

The group is of **fully accessible type** if the above condition is satisfied with $\beta = w(S)$. We see then that Γ is of accessible type if and only if $w(\tilde{H}) > 0$ and is of fully accessible type if and only if $w(\tilde{H}) = w(S)$. Summarizing some of these results we have

Theorem 6.2.7   For a discrete group Γ the following are equivalent.

1. Γ is of fully accessible type.

2. $w(H) = 0$.

3. $\{\xi\colon |\mathsf{h}'(\xi)| < 1 \text{ all } \gamma \in \Gamma - I\}$ is a fundamental region for Γ on S.

4. $\{|\mathsf{h}'(\xi)|\colon \gamma \in \Gamma\}$ accumulates only at zero for almost every $\xi \in S$.

5. $\sum\limits_{\gamma \in \Gamma} |\mathsf{h}'(\xi)|^{n-1} < \infty$ for almost every $\xi \in S$.

We conclude this section by mentioning two examples. The first was given by Hopf [Hopf, 1939 p.275], who constructed a Fuchsian group of the first kind for which ∂D_0 meets S in a set of positive measure. This group then is topometric transitive (theorem 6.2.1) but is not ergodic since it is not conservative (theorem 6.2.5). Hopf's construction will generalize easily to higher dimensions.

The second example (again in dimension 2) is due to Pommerenke [Pommerenke, 1976] who finds a Fuchsian group for which the Dirichlet set D has zero measure and for which there exists a non-constant bounded automorphic function. Since D has zero measure it follows that H has full measure (theorem 2.6.6) and theorem 6.2.4 shows that Γ is conservative on S. The real part of the automorphic function is a bounded, non-constant, Γ-invariant function harmonic in B. By theorem 6.2.2, Γ is not ergodic on S.

6.3 Action on $S \times S$

We start by characterizing regional transitivity.

Theorem 6.3.1   The discrete group Γ is regionally transitive in its action on $S \times S$ if and only if Γ is of the first kind.

Proof. If Γ is of the first kind then, by theorem 2.2.2, there exists a line transitive point, say ξ, in S. The images of any geodesic ending at ξ are dense in the set of all geodesics and so in particular $\Gamma(\xi,-\xi)$ is dense in $S \times S$. This proves that Γ is regionally transitive on $S \times S$.

For the converse, if there exists a point (ξ,η) of $S \times S$ whose Γ-images are dense in $S \times S$ then either ξ or η is a line transitive point. Thus $T_l \neq \varnothing$ and by theorem 2.3.3, Γ is of the first kind. $\square$

The following result, due to Sullivan [Sullivan, 1981], characterizes the conservative and dissipative pieces of the group action on $S \times S$.

Theorem 6.3.2   For the discrete group Γ acting on $S \times S$ the dissipative piece of the action is precisely the set $\tilde{C} \times \tilde{C}$ where $\tilde{C}$ denotes the complement of C in

S. Furthermore the set

$$\{(\xi,\eta)\in S\times S:\, |h'(\xi)||h'(\eta)| < 1 \ \text{for all}\ \gamma\in\Gamma - I\}$$

is a fundamental region for the group action on $\tilde{C}\times\tilde{C}$ — provided the origin is not an elliptic fixed point.

Proof. Suppose (ξ,η), $\xi\neq\eta$, belongs to $\tilde{C}\times\tilde{C}$ and for $\gamma\in\Gamma$ we note that

$$|h'(\xi)||h'(\eta)| = \frac{(1 - |h^{-1}(0)|^2)^2}{|\xi - \gamma^{-1}(0)|^2\,|\eta - \gamma^{-1}(0)|^2} \qquad (6.3.1)$$

from theorem 1.3.4. Since ξ,η are not conical limit points the two sets of reals

$$\left\{\frac{1 - |h^{-1}(0)|^2}{|\xi - \gamma^{-1}(0)|} :\, \gamma\in\Gamma\right\} \quad\text{and}\quad \left\{\frac{1 - |h^{-1}(0)|^2}{|\eta - \gamma^{-1}(0)|} :\, \gamma\in\Gamma\right\} \qquad (6.3.2)$$

each accumulate only at zero. Now for any γ, either $|\xi - \gamma^{-1}(0)|\geq 1/2|\xi - \eta|$ or $|\eta - \gamma^{-1}(0)|\geq 1/2|\xi - \eta|$ and so from (6.3.1)

$$|h'(\xi)||h'(\eta)| \leq \frac{4}{|\xi - \eta|^2}\, Max\left[\left(\frac{1 - |h^{-1}(0)|}{|\xi - \gamma^{-1}(0)|}\right)^2 , \left(\frac{1 - |h^{-1}(0)|}{|\eta - \gamma^{-1}(0)|}\right)^2\right].$$

From (6.3.2) we note that there exists $\gamma_0\in\Gamma$ for which the quantity $|h'(\xi)||h'(\eta)|$ has a maximum. The same γ_0 maximizes $\{|h(\xi) - \gamma(\eta)|:\gamma\in\Gamma\}$. In other words, with $\xi_0 = \gamma_0(\xi)$ and $\eta_0 = \gamma_0(\eta)$,

$$|\xi_0 - \eta_0|\geq |h(\xi_0) - \gamma(\eta_0)|$$

for all $\gamma\in\Gamma$ and this is equivalent to $|h'(\xi_0)||h'(\eta_0)|\leq 1$ — by 1.3.2. We write

$$\Delta = \{(\xi,\eta)\in S\times S:\, |h'(\xi)||h'(\eta)| < 1 \ \textit{for all}\ \gamma\in\Gamma - I\}$$

and note that if (ξ,η) and (ξ_0,η_0) are equivalent then they cannot both be in Δ for that would imply $|\xi - \eta| > |\xi_0 - \eta_0|$ and $|\xi_0 - \eta_0| > |\xi - \eta|$. Our construction above has shown that every (ξ,η) is equivalent to a (ξ_0,η_0) except in the following cases:

1. $\xi = \eta$

2. $\xi\in C$ or $\eta\in C$

3. $|h'(\xi)||h'(\eta)| = 1$ for some $\gamma\in\Gamma - I$.

Clearly (1) and (3) are true only on a set of measure zero and we have proved that Δ is a fundamental region for the group action on $\tilde{C}\times\tilde{C}$. Now suppose $W\subset S\times S$ is a wandering set. Clearly

$$\sum_{\gamma \in \Gamma} \text{measure} \left(\gamma(W) \right) < \infty$$

and so

$$\int_W \sum_{\gamma \in \Gamma} \left(|h'(\xi)| |h'(\eta)| \right)^{n-1} dw(\xi) dw(\eta) < \infty.$$

It follows that for almost every $(\xi, \eta) \in W$,

$$\sum_{\gamma \in \Gamma} \left(|h'(\xi)| |h'(\eta)| \right)^{n-1} < \infty. \tag{6.3.3}$$

Now if either ξ or η is a conical limit point then we may use (6.3.1) and observe that there exists $\epsilon > 0$ and a sequence $\{\gamma_n\} \subset \Gamma$ on which $|h'(\xi)| |h'(\eta)| > \epsilon$. Thus, in view of (6.3.3), any wandering set W must be a subset of $\tilde{C} \times \tilde{C}$. This completes the proof of the theorem. $\square$

Collecting some results from this section and the previous chapter we have the next result.

Theorem 6.3.3   Let Γ be a discrete group acting in B. The following are equivalent.

1. Γ diverges at the exponent $n-1$.

2. $w(C) = w(S)$.

3. Γ is conservative on $S \times S$.

4. $M(\Gamma)$ has no Green's function.

5. The harmonic measure of the ideal boundary of $M(\Gamma)$ is identically zero.

Proof. The equivalence of (1),(4), and (5) is theorem 5.2.1. The equivalence of (2) and (3) follows immediately from theorem 6.3.2 above. The fact that (2) implies (1) is theorem 2.4.4 and so it remains only to prove that (1) implies (2). Suppose that Γ diverges at the exponent $n-1$. Note from theorem 2.4.6 that $w(C) > 0$ and so, since C is a Γ-invariant subset of S, $w(C) = w(S)$ (Γ is ergodic on S by theorem 6.2.3). $\square$

In his 1939 treatise [Hopf, 1939 p.281] Hopf related the ergodicity of Γ on $S \times S$ to the existence or otherwise of bounded two variable functions, harmonic in each variable, which are invariant under Γ (in dimension 2). The results generalize to higher dimensions when we consider h.h functions. We outline the ideas below.

Suppose Γ is not ergodic on $S \times S$ and let A be a measurable Γ-invariant subset of $S \times S$ which is of positive measure and whose complement has positive measure. Writing 1_A for the characteristic function of A we define a function u on $B \times B$ by

$$u(x,y) = \frac{1}{w(S)^2} \int_{S \times S} \left[\frac{(1 - |x|^2)}{|x - \xi|^2} \frac{(1 - |y|^2)}{|y - \eta|^2} \right]^{n-1} 1_A(\xi,\eta) dw(\xi) dw(\eta).$$

It is not difficult to show that this function is Γ-invariant, h.h in each variable and bounded. It can be shown (see [Tsuji, 1959 p.140] for the proof in two dimensions) that $u(x,y)$ has radial limits almost everywhere equal to $1_A(\xi,\eta)$ — thus $u(x,y)$ is non constant.

Now suppose there exists a Γ-invariant, bounded, non constant function in $B \times B$ which is h.h in each variable. Call this function u. It can be shown (see [Tsuji, 1959 p.142] for the proof in two dimensions) that u has radial limits (say $u(\xi,\eta)$ almost everywhere and that

$$u(x,y) = \frac{1}{w(S)^2} \int_{S \times S} \left[\frac{(1 - |x|^2)}{|x - \xi|^2} \frac{(1 - |y|^2)}{|y - \eta|^2} \right]^{n-1} u(\xi,\eta) dw(\xi) dw(\eta).$$

Since u is non constant in $B \times B$, u is not constant almost everywhere on $S \times S$, it is also Γ-invariant and so for any α the set

$$U_\alpha = \{(\xi,\eta): u(\xi,\eta) > \alpha\}$$

is a measurable, Γ-invariant subset of $S \times S$. Clearly for some α both U_α and its complement have positive measure. Thus Γ is not ergodic on $S \times S$ and we have proved the following result.

Theorem 6.3.4   Let Γ be a discrete group acting in B then Γ is ergodic on $S \times S$ if and only if every Γ-invariant, bounded function on $B \times B$ which is h.h in each variable reduces to a constant.

It is clear that if Γ is ergodic on $S \times S$ then it must also be conservative in its action. Thus ergodicity implies all of the five properties given in theorem 6.3.3. Perhaps the most important result in the theory is to the effect that ergodicity is equivalent to these five properties. This result, due to Hopf [Hopf, 1939 p.280] and Tsuji [Tsuji, 1959 p.530] in dimension two, and to Sullivan [Sullivan, 1981] in higher dimensions, is stated below. We will not be in a position to prove this result for some time yet.

Theorem 6.3.5   Let Γ be a discrete group acting in B. Γ is ergodic on $S \times S$ if and only if Γ diverges at the exponent $n-1$.

Considering theorem 6.3.3 we note that if the quotient space $M(\Gamma)$ does not have a Green's function then Γ diverges at the exponent $n-1$. This in turn implies that Γ is ergodic on S (theorem 6.2.3) and so $M(\Gamma)$ does not support a bounded, non constant, h.h function. Using the terminology of Riemann surface theory we have arrived at the generalization of Myrberg's theorem which states that $O_G \subset O_{HB}$. The opposite inclusion is known to be false (for Riemann surfaces — see [Ahlfors and Sario, 1960 p.256]) and this shows in general that a group Γ can be ergodic in its action on S but not ergodic on $S \times S$.

How about topometric transitivity? From the proof of theorem 6.3.1 it is easily seen that Γ is topometric transitive on $S \times S$ if and only if $w(T_l) = w(S)$ — thus topometric transitivity certainly implies that $w(C) = w(S)$ and hence divergence at the exponent $n-1$. From theorem 6.3.5 we see that the converse is also true. Putting these results together with theorems 6.3.3, 6.3.4, and 6.3.5 we have the following.

Theorem 6.3.6   Let Γ be a discrete group acting in B. The following are equivalent.

- Γ diverges at the exponent $n-1$.

- $w(C) = w(S)$.

- Γ is conservative on $S \times S$.

- $M(\Gamma)$ has no Green's function.

- The harmonic measure of the ideal boundary of $M(\Gamma)$ is identically zero.

- Γ is ergodic on $S \times S$.

- Every Γ-invariant, bounded function on $B \times B$ which is h.h in each variable reduces to a constant.

- Γ is topometric transitive on $S \times S$.

6.4 Action on Other Products

To investigate the action of Γ on $S \times S \times S$ we need the following lemma. The proof is straightforward and is omitted.

Lemma 6.4.1   Suppose Γ is a discrete group acting in B and that ξ, η are distinct points of S. Suppose further that on a sequence $\{\gamma_n\}$ of distinct transforms of Γ $\gamma_n(\xi)$ converges to α and that $\gamma_n(\eta)$ converges to β then, possibly on a subsequence, $\gamma_n(0)$ converges either to α or to β.

In dimension two Tsuji [Tsuji, 1959 p.541] showed that any discrete group fails to be ergodic in its action on $S \times S \times S$. We can prove much more than this.

Theorem 6.4.2   Let Γ be a discrete group acting in B then Γ is not regionally transitive on $S \times S \times S$.

Proof. Suppose z_1, z_2, z_3 are three distinct points of S and $\{\gamma_n\}$ is a sequence of transforms of Γ. If $\gamma_n(z_1), \gamma_n(z_2), \gamma_n(z_3)$ converge to α, β, δ respectively then we see from lemma 6.4.1 that α, β, δ cannot be distinct. Thus no point of $S \times S \times S$ has a dense Γ orbit. $\square$

In view of this the following result is not too unexpected.

Theorem 6.4.3   If Γ is a discrete group acting in B then Γ is completely dissipative in its action on $S \times S \times S$. Furthermore the set

$$\{(z_1, z_2, z_3) \in S \times S \times S : |\gamma'(z_1)||\gamma'(z_2)||\gamma'(z_3)| < 1 \text{ for all } \gamma \in \Gamma - I\}$$

is a fundamental region for the group action on $S \times S \times S$ — provided the origin is not an elliptic fixed point.

Proof. For any triple $(z_1, z_2, z_3) \in S \times S \times S$ the set

$$\{|\gamma(z_1) - \gamma(z_2)||\gamma(z_1 - \gamma(z_3)||\gamma(z_2) - \gamma(z_3)| : \gamma \in \Gamma\}$$

is a countable set of non negative reals which, from the proof of theorem 6.4.2, accumulates only at zero. Thus the set has a finite attained maximum. Now from (1.3.2) we see that the set A defined by

$$A = \{(z_1, z_2, z_3) : |\gamma(z_1) - \gamma(z_2)||\gamma(z_1) - \gamma(z_3)||\gamma(z_2) - \gamma(z_3)|$$

$$< |z_1 - z_2||z_1 - z_3||z_2 - z_3| \text{ for all } \gamma \in \Gamma - I\}$$

is the same as the set

$$\{(z_1, z_2, z_3) \in S \times S \times S : |\gamma'(z_1)||\gamma'(z_2)||\gamma'(z_3)| < 1 \text{ for all } \gamma \in \Gamma - I\}.$$

Our remarks above show that each triple (z_1, z_2, z_3) of $S \times S \times S$ has a group image in A — unless it belongs to the set of zero measure for which $|\gamma'(z_1)||\gamma'(z_2)||\gamma'(z_3)| = 1$ for some $\gamma \in \Gamma - I$. Clearly, from the definition above, A is measurable and contains no two Γ-equivalent points. This completes the proof of the theorem. $\square$

Corollary 6.4.4   If Γ is a discrete group acting in B then the series

$$\sum_{\gamma \in \Gamma} \left(|\gamma'(z_1)||\gamma'(z_2)||\gamma'(z_3)| \right)^{n-1}$$

converges for almost every triple $(z_1, z_2, z_3) \in S \times S \times S$.

Proof. The set A defined in the proof of theorem 6.4.3 is of positive measure and has Γ images no two of which overlap thus

$$\sum_{\gamma \in \Gamma} \text{measure } (\gamma(A)) \; < \; \infty$$

and from this follows the convergence of the integral

$$\int_A \sum_{\gamma \in \Gamma} \left(|\gamma'(z_1)||\gamma'(z_2)||\gamma'(z_3)| \right)^{n-1} dw(z_1)dw(z_2)dw(z_3).$$

The series $\sum_{\gamma \in \Gamma} \left(|\gamma'(z_1)||\gamma'(z_2)||\gamma'(z_3)| \right)^{n-1}$ thus converges almost everywhere in A and the corollary follows immediately. $\square$

We now consider the question of ergodicity. Either one of the two theorems 6.4.2.or 6.4.3 shows that no discrete group is ergodic in its action on $S \times S \times S$. However, it is possible to show, in the special case $n = 2$ that even the full Moebius group is not ergodic in its action on $S \times S \times S$.

Theorem 6.4.5   In dimension 2 there exists a subset A of $S \times S \times S$ which has positive measure, whose complement has positive measure and which is invariant under the full Moebius group.

Proof. Define A as the set of triples (z_1, z_2, z_3) with z_i distinct and positively oriented around the unit circle. Similarly, A^* is the set of distinct triples negatively oriented around the circle. A and A^* are disjoint measurable subsets of $S \times S \times S$ each of which has positive measure (allow each z_j to move over a suitably small arc) and each of which is invariant under all Moebius transforms preserving the unit disc. $\square$

Theorem 6.4.5 certainly fails in higher dimensions. To see this recall that if (z_1, z_2, z_3) , (w_1, w_2, w_3) are two points of $\hat{C} \times \hat{C} \times \hat{C}$ there exists a Moebius transform γ of $\hat{C}$ such that $\gamma(z_i) = w_i$, $i = 1, 2, 3$. Now the Poincaré extension $\hat{\gamma}$ of γ is a Moebius transform of R^3 which preserves the upper half-space and such that $\hat{\gamma}(z_i) = w_i$, $i = 1, 2, 3$. Conjugating to the unit ball we see that in dimension three the full Moebius group acts ergodically on $S \times S \times S$ and so no set A as described in theorem 6.4.5 can exist.

Theorem 6.4.6   In any dimension the following are true.

1. The full Moebius group is not regionally transitive on $S \times S \times S \times S$

2. There is a subset A of $S \times S \times S \times S$ which is of positive measure, whose complement is of positive measure, and which is invariant under the full Moebius group.

Proof. (1) follows trivially from the invariance and continuity of the cross ratio defined in section 1.3. To prove (2) simply define

$$A \; = \; \{(z_1, z_2, z_3, z_4) : z_i \neq z_j, j \neq i \; ; \; z_j \in S, \; j \; = 1,2,3,4 \text{ and } |z_1, z_2, z_3, z_4| < 1\}$$

and the result follows immediately. □

CHAPTER 7

Elementary Ergodic Theory

7.1 Introduction

In this chapter we develop, in a general setting, the ergodic theorems which we will need in order to study certain flows on the quotient space of a discrete group. In this introductory section we state, without proof, three classical ergodic theorems and use these in the next section to develop the particular results needed for our applications. The proofs of the results given in this section can be found in many texts on ergodic theory — for example [Cornfeld, Fomin, and Sinai, 1982], [Nemytskii and Stepanov, 1960], and [Walters, 1982]. The section concludes with the statement and proof of a result of Hopf.

The general setting is a sigma finite measure space (Ω, B, μ) on which acts a measure preserving transform T. For a real valued function f defined on Ω we consider the sequence f_n defined by

$$f_n(x) = \frac{1}{n} \sum_{j=0}^{n-1} f(T^j(x)). \qquad (7.1.1)$$

We begin with Von Neumann's mean ergodic theorem — see [Walters, 1982 p.36].

Theorem 7.1.1   Let $1 \leq p \leq \infty$, if $f \in L^p(\Omega)$ there exists $f^* \in L^p(\Omega)$ with $f^*(x) = f^*(T(x))$ for almost every $x \in \Omega$ and

$$\|f_n - f^*\|_p \to 0 \quad as \quad n \to \infty.$$

The next result is Birkhoff's individual ergodic theorem — see [Walters, 1982 p.34].

Theorem 7.1.2   If $f \in L^1(\Omega)$ there exists $f^* \in L^1(\Omega)$ with $f^*(x) = f^*(T(x))$ for almost every $x \in \Omega$ and

$$f_n(x) \rightarrow f^*(x) \quad as \ n \rightarrow \infty$$

for almost every $x \in \Omega$. Further, if $\mu(\Omega) < \infty$, then

$$\int_\Omega f^* d\mu = \int_\Omega f d\mu.$$

Our third result, which is used in the proof of Birkhoff's theorem, is called the maximal ergodic theorem — see [Walters, 1982 p.37].

Theorem 7.1.3   If $f \in L^1(\Omega)$, let E be the set of points $x \in \Omega$ for which at least one of the sums

$$f(x) + f(T(x)) + f(T^2(x)) + \cdots + f(T^n(x))$$

is non negative, then $\int_E f d\mu \geq 0$.

An important generalization of the Birkhoff theorem was derived by E. Hopf [Hopf, 1937]. We conclude this section by stating and proving this result.

Theorem 7.1.4   Suppose $f, g \in L^1(\Omega)$ with $g > 0$ and that for almost every $x \in \Omega$, $\lim_{n \rightarrow \infty} \sum_{j=0}^{n-1} g(T^j(x)) = \infty$. Then the sequence

$$\left[\sum_{j=0}^{n-1} f(T^j(x)) \right] \left[\sum_{j=0}^{n-1} g(T^j(x)) \right]^{-1}$$

converges almost everywhere to a measurable function $\phi(x)$. The function ϕ is invariant under T, the function $g\phi$ is integrable, and for every bounded measurable function h invariant under T,

$$\int_\Omega g\phi h \ d\mu = \int_\Omega fh \ d\mu.$$

Note that if $\mu(\Omega) < \infty$ we may take $g \equiv h \equiv 1$ in this theorem and recover Birkhoff's theorem (in the finite measure case).

Proof. For real numbers a, b with $a < b$ let $Y(a, b)$ be the set of $x \in \Omega$ with

$$\liminf_{n \rightarrow \infty} \frac{\sum_{j=0}^{n-1} f(T^j(x))}{\sum_{j=0}^{n-1} g(T^j(x))} < a < b < \limsup_{n \rightarrow \infty} \frac{\sum_{j=0}^{n-1} f(T^j(x))}{\sum_{j=0}^{n-1} g(T^j(x))}.$$

Clearly $Y(a, b)$ is measurable and, considering

$$\frac{\sum\limits_{j=1}^{n} f(T^j(x))}{\sum\limits_{j=1}^{n} g(T^j(x))} = \frac{\dfrac{\sum\limits_{j=0}^{n-1} f(T^j(x))}{\sum\limits_{j=0}^{n-1} g(T^j(x))} + \dfrac{f(T^n(x))}{\sum\limits_{j=0}^{n-1} g(T^j(x))} - \dfrac{f(x)}{\sum\limits_{j=0}^{n-1} g(T^j(x))}}{1 + \dfrac{g(T^n(x))}{\sum\limits_{j=0}^{n-1} g(T^j(x))} - \dfrac{g(x)}{\sum\limits_{j=0}^{n-1} g(T^j(x))}}$$

we see that $Y(a,b)$ is T-invariant. Now note that

$$\sum_{j=0}^{n-1} [\, f(T^j(x)) - bg(T^j(x))\,] \geq 0$$

if and only if

$$\frac{\sum\limits_{j=0}^{n-1} f(T^j(x))}{\sum\limits_{j=0}^{n-1} g(T^j(x))} \geq b$$

which is true almost everywhere in $Y(a,b)$ for some n. We appeal to theorem 7.1.3 with Ω replaced by $Y(a,b)$ and f replaced by $f - bg$ to see that

$$\int_{Y(a,b)} f - bg \, d\mu \geq 0.$$

Similarly,

$$\int_{Y(a,b)} ag - f \, d\mu \geq 0$$

and so

$$(a - b) \int_{Y(a,b)} g \, d\mu \geq 0.$$

Since $a < b$ and $g > 0$ we have $\mu[Y(a,b)] = 0$. Applying this to all pairs of rationals with $a < b$ we have proved that the limit

$$\lim_{n \to \infty} \frac{\sum\limits_{j=0}^{n-1} f(T^j(x))}{\sum\limits_{j=0}^{n-1} g(T^j(x))}$$

exists almost everywhere. The limit function $\phi(x)$ is clearly measurable and is T-invariant for the same reason that $Y(a,b)$ is.

If $\phi \geq a$ we can apply the maximal ergodic theorem to $f - g(a - \epsilon)$ and obtain $\int_\Omega f \, d\mu \geq a \int_\Omega g \, d\mu$. Similarly, if $\phi \leq b$ everywhere then

$b \int_\Omega g \, d\mu \geq \int_\Omega f \, d\mu$. Now for integers k,n define

$$X(k,n) = \{x : k/2^n \leq \phi(x) \leq (k+1)/2^n\}$$

which is a measurable, T-invariant subset of Ω. Replacing Ω by $X(k,n)$ in the maximal ergodic theorem we see from our remarks above that

$$\frac{k+1}{2^n} \int_{X(k,n)} g \, d\mu \geq \int_{X(k,n)} f \, d\mu \geq \frac{k}{2^n} \int_{X(k,n)} g \, d\mu.$$

Trivially,

$$\frac{k+1}{2^n} \int_{X(k,n)} g \, d\mu \geq \int_{X(k,n)} g\phi \, d\mu \geq \frac{k}{2^n} \int_{X(k,n)} g \, d\mu$$

and so

$$\frac{-1}{2^n} \int_{X(k,n)} g \, d\mu \leq \int_{X(k,n)} f \, d\mu - \int_{X(k,n)} g\phi \, d\mu \leq \frac{1}{2^n} \int_{X(k,n)} g \, d\mu.$$

Summing over k we obtain

$$\left| \int_\Omega f \, d\mu - \int_\Omega g\phi \, d\mu \right| \leq \frac{1}{2^n} \int_\Omega g \, d\mu,$$

and letting $n \rightarrow \infty$ we have proved that

$$\int_\Omega f \, d\mu = \int_\Omega g\phi \, d\mu. \tag{7.1.2}$$

Now with h as given in the theorem, we simply set $f_1 = hf$, note that $\sum_{j=0}^{n-1} f_1(T^j(x)) = h(x) \sum_{j=0}^{n-1} f(T^j(x))$ and appeal to (7.1.2) to obtain

$$\int_\Omega fh \, d\mu = \int_\Omega g\phi h \, d\mu$$

as required. $\square$

7.2 The Continuous Case

In the previous section we considered the action of powers of a measure preserving transformation on a measure space. We now consider the situation of a continuous flow defined on a space. The general setting is a separable, complete metric space (Ω,d) which is equipped with a σ-finite measure μ on its Borel subsets. We further assume that Ω is locally compact and that the measure of any compact subset of Ω is finite.

A **flow** on Ω is a map $\pi : \Omega \times R \rightarrow \Omega$ with the following four properties:

1. $\pi(x,0) = x$ for all $x \in \Omega$.

2. $\pi(\pi(x,s),t) = \pi(x,t+s)$ all $x \in \Omega$ and all t,s real.

3. π is continuous.

4. For any $t \in R$ and A a measurable subset of Ω, $\pi(A,t)$ is measurable and $\mu(\pi(A,t)) = \mu(A)$.

We will usually write

$$\pi(x,s) = T_s(x)$$

and make several definitions as follows.

Given $x \in \Omega$ the **trajectory** of x, written C_x, is defined by $C_x = \{T_t(x) : t \in R\}$. If $A \subset \Omega$ and $T_t(A) = A$ for all $t \in R$ then we say that A is **invariant** under the flow. Given $x \in \Omega$ and $E \subset \Omega$ then x is **recurrent** with respect to E if $T_t(x) \in E$ for a sequence of t tending to ∞. Our first result is the Poincaré recurrence theorem — see [Walters, 1982 p.26].

Theorem 7.2.1   Suppose $\mu(\Omega) < \infty$ and $A \subset \Omega$ is measurable with $\mu(A) > 0$, then x is recurrent with respect to A for almost all $x \in A$.

Proof.   Define $T : \Omega \to \Omega$ by $T(x) = T_1(x)$ and let $A \subset \Omega$ be measurable. Define

$$F = A \cap T^{-1}(X - A) \cap T^{-2}(X - A) \cap \cdots$$

which is clearly a measurable subset of Ω and comprises those points of A which never return to A under iterates of T. Now if $x \in F$ then none of the points $T(x), T^2(x),...$ belong to F and so F is disjoint from $T^n(F)$ for all positive n. It follows that the sets $F, T(F),...$ are pairwise disjoint since $T^n(F) \cap T^{n+k}(F) = T^n(F \cap T^k(F))$. Since T is measure preserving and X has finite measure we see that $\mu(F) = 0$. For integer n let F_n be the set of points of A which never return to A under iterates of T^n — then, by our argument above, $\mu(F_n) = 0$. If $x \in A - [F_1 \cup F_2 \cup ...]$ then $T^n(x) \in A$ for some positive integer n since $x \in A - F_1$. Similarly, since $x \in A - F_n$, it follows that $T^{k_n}(x) \in A$ for some positive integer k. We have shown that for almost all $x \in A$, $T^n(x) \in A$ for infinitely many values of n. Noting that $T^n = T_n$ the theorem is proved. □

The flow T_s on Ω is said to be **regionally transitive** if for some $x \in \Omega$, C_x is dense in Ω.

Theorem 7.2.2   The following are equivalent:

1. T_s is regionally transitive.

2. Every open set in Ω which is invariant under T_s is everywhere dense in Ω.

3. If A, A^* are any open sets in Ω there exists t such that $T_t(A) \bigcap A^* \neq \varnothing$.

Proof. That (3) implies (2) is trivial. To show that (1) implies (3) we suppose an x given with C_x dense in Ω. For some s,t real we have $T_t(x) \in A$, $T_s(x) \in A^*$ and so $T_{s-t}(A) \bigcap A^* \neq \varnothing$. It remains only to show that (2) implies (1). Let $\{U_i\}$ be a countable base for the topology on Ω. C_x is not dense in Ω if and only if for some n, $C_x \bigcap U_n = \varnothing$ which means that $C_x \subset \{\Omega - U_n\}$ and this is equivalent to saying that there exists n with

$$x \in \bigcap_{s \in R} T_s \{\Omega - U_n\}.$$

This happens if and only if

$$x \in \bigcup_{n=1}^{\infty} \bigcap_{s \in R} T_s \{\Omega - U_n\}.$$

We write $A_n = \bigcap_{s \in R} T_s \{\Omega - U_n\}$ and note that $\tilde{A}_n = \bigcup_{s \in R} T_s(U_n)$ is open and flow invariant and so by property (2) is everywhere dense in Ω. Thus A_n is nowhere dense. It follows from our remarks above that for some $x \in \Omega$, C_x is dense in Ω, since otherwise Ω (a complete metric space) is a countable union of nowhere dense subsets — a contradiction with the Baire category theorem. $\square$

A much stronger notion is topometric transitivity. The flow T_s on Ω is **topometric transitive** if C_x is dense in Ω for almost all $x \in \Omega$.

Theorem 7.2.3   The following are equivalent:

1. T_s is topometric transitive.

2. If M is a measurable subset of Ω which is of positive measure and invariant under T_s then M is everywhere dense in Ω.

3. If M is a measurable subset of Ω which is of positive measure and D is an open set in Ω then for some t, $T_t(M) \bigcap D \neq \varnothing$.

Proof. That (3) implies (2) is easy. To show that (1) implies (3), given M as in (3), find $x \in M$ with C_x dense in Ω then there exists t such that $T_t(M) \bigcap D$ contains $T_t(x)$ and so is non empty.

It remains to show that (2) implies (1). Let $\{U_i\}$ be a countable base for the topology on Ω. We define A_n as in the proof of theorem 7.2.2 and noting that $\tilde{A}_n$ is open and T_t-invariant we see that A_n is measurable and T_t-invariant. If $\mu(A_n) > 0$ then by (2), $A_n \bigcap U_n \neq \varnothing$. But $U_n \subset \tilde{A}_n$ so we must have $\mu(A_n) = 0$

and it follows that $\mu\{x : \overline{C}_x \neq \Omega\} = 0$ as required. $\square$

Before proceeding to deeper ergodic properties we consider, for functions $f \in L^1(\Omega)$, the existence of integrals of the type

$$\int_0^t f(T_s(x))ds$$

Theorem 7.2.4   If $f \in L^1(\Omega)$ then $f(T_s(x))$ is measurable as a function on $\Omega \times R$ with the product measure.

Proof. If $A \subset \Omega$ define the "tube"

$$T_R(A) = \{ (x,t) : T_t(x) \in A \}.$$

Note that if $O \subset \Omega$ is open then $T_R(O)$ is open, and hence measurable in $\Omega \times R$. Now suppose $A \subset \Omega$ is a G_δ set, $A = \bigcap_{n=1}^{\infty} O_n$ then

$$T_R(A) = \{(x,t) : T_t(x) \in A\} = \{(x,t) : T_t(x) \in \bigcap_{n=1}^{\infty} O_n\}$$

$$= \bigcap_{n=1}^{\infty} \{(x,t) : T_t(x) \in O_n\}$$

$$= \bigcap_{n=1}^{\infty} T_R(O_n)$$

so $T_R(A)$ is a G_δ set in $\Omega \times R$ and hence is measurable.

Now suppose that E is a measurable set in Ω. We set $E = A - N$ where A is a G_δ set and N is a null set. Clearly $T_R(E) = T_R(A) - T_R(N)$ and we show next that $T_R(N)$ is a null set. For positive integer M let $T_M(N) = \{(x,t) : T_t(x) \in N, |t| < M\}$ then

$$T_R(N) = \bigcup_{M=1}^{\infty} T_M(N).$$

For $\epsilon > 0$ let $O \supset N$ be an open set with $\mu(O) < \epsilon$. If ν denotes the product measure on $\Omega \times R$,

$$\nu(T_M(O)) = \int_{\Omega \times R} 1_{T_M(O)} \, d\nu = \int_{-M}^{M} \int_\Omega 1_{T_M(O)}(x,t) d\mu \, dt$$

by Fubini's theorem. But T_t is measure preserving and so the inner integral is $\mu(O)$ and $\nu(T_M(O)) < 2M\epsilon$. Letting $\epsilon \to 0$ we see that $T_M(N)$ is a null set and consequently that $T_R(N)$ is a null set.

We have shown that if $E \subset \Omega$ is μ - measurable then $T_R(E)$ is ν measurable. If f is the characteristic function of E then $f(T_t(x))$ is measurable in $\Omega \times R$. From this the theorem follows easily since any measurable f is a limit of countable sums of characteristic functions. $\square$

We next give the Von Neumann and Birkhoff Theorems for flows.

Theorem 7.2.5   If $f \in L^2(\Omega)$ there exists $f^* \in L^2(\Omega)$ with $f^* = f^* o T_s$ for almost every $x \in \Omega$ and

$$\left\| f^* - \frac{1}{S} \int_0^S f(T_s(x)) ds \right\|_2 \longrightarrow 0$$

as $S \longrightarrow \infty$.

Proof.   Set $T = T_1$ and define $F(x) = \int_0^1 f(T_s(x)) ds$. From Hölder's inequality on the interval $[0,1]$, applied to the functions $f(T_s(x))$ and 1, we know that

$$\int_0^1 |f(T_s(x))| ds \leq \left(\int_0^1 f(T_s(x))^2 ds \right)^{1/2}$$

from which it follows that

$$\int_\Omega F(x)^2 d\mu = \int_\Omega \left(\int_0^1 f(T_s(x)) ds \right)^2 d\mu$$

$$\leq \int_\Omega \int_0^1 f(T_s(x))^2 ds \, d\mu$$

$$= \int_0^1 \int_\Omega f(T_s(x))^2 d\mu \, ds$$

using Fubini's theorem. However, T_s is measure preserving and $f \in L^2(\Omega)$ and it follows that $F \in L^2(\Omega)$. We next note that

$$\int_j^{j+1} f(T_s(x)) ds = \int_0^1 f(T_{u+j}(x)) du$$

$$= \int_0^1 f(T^j(T_u(x))) du$$

$$= F(T^j(x))$$

and so

$$\sum_{j=0}^{n-1} F(T^j(x)) = \int_0^n f(T_s(x)) ds.$$

Applying the discrete theorem (theorem 7.1.1) to the function F and the transform T we find the existence of a function $f^* \in L^2(\Omega)$ which is invariant under T and such that

$$\left\| f^*(x) - \frac{1}{n} \int_0^n f(T_s(x))\, ds \right\|_2 \to 0 \quad \text{as } n \to \infty. \qquad (7.2.1)$$

Given that the limit

$$\lim_{u \to \infty} \frac{1}{u} \int_0^u f(T_s(x))ds$$

is known to exist almost everywhere we deduce from (7.2.1) that it is equal almost everywhere to f^* and is the L^2 limit of $\frac{1}{u} \int_0^u f(T_s(x))ds$. This proves the theorem. $\square$

In an entirely similar fashion one may prove the continuous version of the individual ergodic theorem.

Theorem 7.2.6   If T_s is a flow on the space Ω and if $f \in L^1(\Omega)$ then the limit

$$f^*(x) \;=\; \lim_{u \to \infty} \frac{1}{u} \int_0^u f(T_s(x))ds$$

exists almost everywhere, is integrable and flow invariant. Further, if $\mu(\Omega) < \infty$ then

$$\int_\Omega f^*\, d\mu \;=\; \int_\Omega f\, d\mu.$$

We also have the continuous version of Hopf's generalization of the Birkhoff theorem.

Theorem 7.2.7   If T_s is a flow on Ω, if $f, g \in L^1(\Omega)$, $g > 0$ and

$$\int_0^u g(T_s(x))\, ds \;\to\; \infty$$

as $u \to \infty$ almost everywhere on Ω then the limit

$$\phi(x) \;=\; \lim_{u \to \infty} \frac{\int_0^u f(T_s(x))\, ds}{\int_0^u g(T_s(x))\, ds}$$

exists almost everywhere. The function ϕ is measurable and flow invariant. The function $g\phi$ is integrable over Ω and, for any bounded, measurable T_s-invariant function h,

$$\int_\Omega g\phi h\, d\mu \;=\; \int_\Omega f h\, d\mu.$$

Proof. As in the proof of theorem 7.2.5 set $T = T_1$ and define

$$F(x) = \int_0^1 f(T_s(x))ds, \qquad G(x) = \int_0^1 g(T_s(x))ds.$$

The theorem follows by applying the discrete result (theorem 7.1.4) to the transform T and the functions F, G. □

Of fundamental importance is the notion of an ergodic flow which is defined as follows. The flow T_s on the space Ω is said to be **ergodic** if, whenever M is a measurable flow invariant subset of Ω, either $\mu(M) = 0$ or $\mu(\tilde{M}) = 0$.

The function ϕ given in the previous theorem is measurable and flow invariant thus, for an ergodic flow, it must be constant almost everywhere and the following result is immediate.

Theorem 7.2.8   If T_t is ergodic and f, g satisfy the hypotheses of theorem 7.2.7 then, for almost all $x \in \Omega$,

$$\lim_{u \to \infty} \frac{\int_0^u f(T_s(x))\,ds}{\int_0^u g(T_s(x))\,ds} = \frac{\int_\Omega f\,d\mu}{\int_\Omega g\,d\mu}.$$

For a space Ω of finite measure we may take $g \equiv 1$ in theorem 7.2.8 and deduce that the **time mean**

$$\lim_{u \to \infty} \frac{1}{u} \int_0^u f(T_s(x))\,ds$$

is equal almost everywhere to the **space mean**

$$\frac{1}{\mu(\Omega)} \int_\Omega f\,d\mu$$

for the function f.

This important result says that an ergodic flow produces a good mixture of the points in the space - for **any** L^1 function f the space mean is reflected ultimately in the time mean along almost every trajectory.

A major concern with this result is that it can only be used if there exists a positive L^1 function g with the property that

$$\int_0^u g(T_s(x))ds \to \infty \quad as \quad u \to \infty$$

almost everywhere on Ω. The existence of such a function (at least on spaces Ω of infinite measure) is far from clear. In order to understand this situation better we need to consider the conservative and dissipative sets associated with a flow. Let T_t be a flow on Ω. We say that x is a **conservative** point for the flow if $T_t(x)$ remains in some compact set F for a sequence of t tending to infinity. Otherwise, if $T_t(x)$ ultimately leaves any compact set (for t large enough), then x is a **dissipative** point. We write C for the conservative set and D for the dissipative set and note that $C \bigcup D = \Omega$. In the finite volume case the conservative set comprises almost all of Ω.

Theorem 7.2.9   If T_t is a flow on a space Ω with $\mu(\Omega) < \infty$ then $\mu(C) = \mu(\Omega)$.

Proof. Since Ω is locally compact we have a countable collection $\{F_n\}$ of compact subsets of Ω with

$$\Omega = \bigcup_{n=1}^{\infty} F_n.$$

Write $F^N = \bigcup_{n=1}^{N} F_n$ and, since $\mu(\Omega) < \infty$, we have $\mu(F^N) \to \mu(\Omega)$ as $N \to \infty$. From theorem 7.2.1, $\mu(C \bigcap F^N) = \mu(F^N)$, and so

$$\mu(C) = \mu(C \bigcap \Omega) = \mu(C \bigcap (\bigcup_{n=1}^{\infty} F^N))$$
$$= \mu \left[\bigcup_{n=1}^{\infty} (C \bigcap F^N) \right]$$
$$\geq \mu(F^N) \text{ for } every \text{ } N$$

and so $\mu(C) = \mu(\Omega)$ as required. □

Concerning the existence of a function g to satisfy the hypotheses of theorem 7.2.7 we have the following result

Theorem 7.2.10   If T_t is a flow on the space Ω then there exists an L^1 positive function g defined on Ω such that

$$\int_0^u g(T_t(x))dt \to \infty \text{ as } u \to \infty$$

for all $x \in C$.

Proof. Since Ω has a countable base and is locally compact we can construct a sequence of compact sets

$$F_0 \subset F_1 \subset F_2 \subset ..., \quad \lim F_N = \Omega.$$

We may assume that for each n there exists ϵ_n such that

$$\{x : d(x, F_n) \leq \epsilon_n\} \subset F_{n+1}.$$

This is an easy but technical result — see [Nemytskii and Stepanov, 1960 p.477]. Let $\mu(F_0) = m_0$, $\mu(F_n - F_{n-1}) = m_n$, $n = 1, 2, ...$ and note that all of these numbers are finite. Clearly we have

$$\Omega = \bigcup_{n=1}^{\infty} (F_n - F_{n-1}) \bigcup F_0$$

and this is a disjoint union. We define $g : \Omega \to R^+$ by

$$g(x) = (2^n m_n)^{-1} \quad \text{if} \quad x \in F_n - F_{n-1}$$
$$g(x) = m_0^{-1} \quad \text{if} \quad x \in F_0.$$

Clearly g is measurable and $\int_\Omega g \, d\mu = 2$. Now suppose $x \in C$, then there exists a compact set F and a sequence $\{t_n\}$ tending to infinity such that $x_n = T_{t_n}(x) \in F$. Now $F \subset F_N$ for some N and we define $\rho = (2^{N+1} m(F_{N+1}))^{-1} > 0$ noting that if $w \in F_{N+1}$, $g(w) \geq \rho$. For each n we let s_n be the real number satisfying $t_n < s_n < t_{n+1}$ and $d(T_{t_n}(x), T_{s_n}(x)) = \epsilon_N$ — if it exists. If such an s_n does not exist then the segment $\{T_t(x) : t_n \leq t \leq t_{n+1}\}$ of the trajectory of x does not leave F_{N+1}. Thus if s_n exists for only finitely many n we will have $T_t(x) \in F_{N+1}$ all $t \geq T$ and trivially

$$\int_0^\infty g(T_t(x)) dt = \infty.$$

If infinitely many s_n exist we claim that

$$\inf_n (s_n - t_n) = \epsilon > 0. \tag{7.2.2}$$

To see this, observe that, on a subsequence if necessary, $x_n \to y$ in F_N, and that $d(T_{t_n}(x_n), T_{s_n}(x_n)) = \epsilon_N$. Thus (7.2.2) must be true by the continuity of the flow. Note that

$$\int_0^\infty g(T_t(x)) dt \geq \sum_{n=1}^{\infty} \int_{t_n}^{s_n} g(T_t(x)) dt \geq \sum_{n=1}^{\infty} \rho \epsilon = \infty$$

and the proof is complete. $\square$

We are now in a position to prove the following result which characterizes ergodicity.

Theorem 7.2.11   For a flow T_t on a space Ω the following are equivalent.

1. T_t is ergodic.

2. If M , M^* are two measurable sets of positive measure then there exists $t > 0$ such that $T_t(M) \bigcap M^* \neq \emptyset$.

3. If $f, h \in L^1(\Omega)$ then, provided $\int_\Omega h \, d\mu \neq 0$,

$$\lim_{u \to \infty} \frac{\int_0^u f(T_t(x))\, dt}{\int_0^u h(T_t(x))\, dt} = \frac{\int_\Omega f\, d\mu}{\int_\Omega h\, d\mu}$$

almost everywhere in Ω.

Proof. We show first that (1) implies (3). If T_t is ergodic then by theorem 7.2.3 it is certainly topometric transitive and it follows from this that the conservative set comprises almost all of Ω. Thus we may find a function g satisfying the hypotheses of theorem 7.2.8 and, from that theorem,

$$\lim_{u \to \infty} \frac{\int_0^u f(T_t(x))\, dt}{\int_0^u h(T_t(x))\, dt} = \lim_{u \to \infty} \frac{\int_0^u f(T_t(x))\, dt}{\int_0^u g(T_t(x))\, dt} \cdot \frac{\int_0^u g(T_t(x))\, dt}{\int_0^u h(T_t(x))\, dt}$$

$$= \frac{\int_\Omega f\, d\mu}{\int_\Omega h\, d\mu}$$

as required. To show that (3) implies (2) we let h be a positive integrable function, we suppose M, M^* are of positive measure and that $T_t(M) \cap M^* = \emptyset$ for all t. If $f = 1_{M^*}$ we have a contradiction to (3) if $x \in M$. Finally we show that (2) implies (1). Suppose A is a measurable flow invariant set with $\mu(A) > 0$ and $\mu(\tilde{A}) > 0$ - this immediately yields a contradiction with (2) if we take $M = A$ and $M^* = \tilde{A}$. $\square$

Corollary 7.2.12   If T_t is ergodic and A, B are two measurable subsets of Ω of finite measure and $\mu(B) \neq 0$ then

$$\lim_{u \to \infty} \frac{\frac{1}{u} \int_0^u 1_A(T_t(x))\, dt}{\frac{1}{u} \int_0^u 1_B(T_t(x))\, dt} = \frac{\mu(A)}{\mu(B)}$$

for almost every $x \in \Omega$.

The last, and strongest, ergodic property we will consider is called mixing. The flow T_t on Ω is said to be **mixing** if for any measurable subsets A, B, C of Ω which are of finite measure and with $\mu(B) \neq 0$ we have

$$\lim_{t \to \infty} \frac{\mu(T_t(A) \cap C)}{\mu(T_t(B) \cap C)} = \frac{\mu(A)}{\mu(B)}.$$

The property we have just described is called "strong mixing" by many authors — [Walters, 1982 p.40] for example. Note that if $\mu(\Omega) < \infty$ then mixing is equivalent to the condition that for A, B measurable subsets of Ω

$$\lim_{t \to \infty} \mu(T_t(A) \cap B) = \frac{\mu(A)\mu(B)}{\mu(\Omega)}.$$

We conclude this section by demonstrating the relation between the various ergodic properties introduced so far in this chapter.

Theorem 7.2.13   If T_t is a flow on the space Ω then each of the following properties implies the next.

1. T_t is mixing.

2. T_t is ergodic.

3. T_t is topometric transitive.

4. T_t is regionally transitive.

Proof. To see that (1) implies (2) we assume (1) and, in the definition of mixing, we let $B = C$ be a T_t-invariant measurable subset of Ω which has positive measure. Thus we obtain

$$\lim_{t \to \infty} \mu(T_t(A) \cap B) = \mu(A).$$

Now let $A \subset \tilde{B}$ be of finite measure. Since $\tilde{B}$ is T_t-invariant it follows that $T_t(A) \subset \tilde{B}$ and we deduce that $\mu(A) = 0$. Thus $\tilde{B}$ contains no subset of positive finite measure and it follows that $\mu(\tilde{B}) = 0$. This proves (2). The remaining implications in the theorem are immediate consequences of theorems 7.2.2, 7.2.3, and 7.2.11.

7.3 Invariant Measures

In the previous section we assumed that the space Ω came equipped with a σ-finite measure μ and we defined the flow so as to guarantee that it preserved the measure of Borel subsets of Ω.

In this section we briefly consider questions concerning the existence and uniqueness of measures preserved by flows. To this end, let us suppose we have a separable complete metric space (Ω, d) which is compact. Suppose further that π is a map from $\Omega \times R$ to Ω satisfying the first three properties of a flow. Namely:

1. $\pi(x,0) = x$ for all $x \in \Omega$.

2. $\pi(\pi(x,s),t) = \pi(x,t+s)$ for all $x \in \Omega$ and all t,s real.

3. π is continuous.

Let us consider the possibility of finding a measure μ on Borel subsets of Ω which is invariant under the flow (we write $T_s(x)$ for $\pi(x,s)$) and such that $\mu(\Omega) = 1$. If we suppose further that T_t is ergodic with respect to the measure μ, then if E is a Borel subset of Ω we may take $f = 1_E$ and $g \equiv 1$ in theorem 7.2.8 to see that for almost every (μ) $x \in \Omega$

$$\mu(E) = \lim_{u \to \infty} \frac{1}{u} \int_0^u 1_E(T_t(x))dt.$$

This suggests a definition for a T_t-invariant probability measure. Given T_t defined on Ω, $x \in \Omega$ and s a positive real number, one can define, for any Borel set E

$$M_{x,s}(E) = s^{-1} \int_0^s 1_E(T_t(x))dt,$$

and this yields a Borel probability measure. However, this measure is not T_t-invariant — one needs to go further and consider limits (in the topology of weak convergence) of such measures as $s \to \infty$. The interested reader is referred to [Kryloff and Bogolioubov, 1937] for a full account (see also [Nemytskii and Stepanov, 1960 pp. 486 - 519] and [Walters, 1960 Chapter 6]).

For a compact space Ω it is always possible to construct in this fashion a T_t-invariant probability measure on Ω. How about uniqueness? The following result gives some information on this point.

Theorem 7.3.1   If μ is a T_t-invariant measure on Ω with respect to which T_t is not ergodic then there exists a T_t-invariant measure ν on Ω and a Borel subset E of Ω such that $\mu(E) > 0$ and $\nu(E) = 0$.

Proof. Since the flow T_t is not ergodic with respect to μ there exists a Borel subset A of Ω which is T_t-invariant with $\mu(A) > 0$ and $\mu(\tilde{A}) > 0$. Define a measure ν by

$$\nu(F) = \mu(A \cap F)$$

for any Borel subset F of Ω and note that

$$\begin{aligned}
\nu(T_t(F)) &= \mu(A \cap T_t(F)) \\
&= \mu(T_t(A) \cap T_t(F)) \\
&= \mu(A \cap F) = \nu(F)
\end{aligned}$$

and so ν is T_t-invariant. Now set $E = \tilde{A}$ and the theorem is proved. $\square$

Flows with precisely one invariant Borel probability measure are called **uniquely ergodic** — we will return to a consideration of such flows in chapter 10.

CHAPTER 8

The Geodesic Flow

8.1 Definition

There is an enormous body of literature on the geodesic flow for manifolds of constant (or even variable) negative curvature. In this chapter we will present the basic results relating to ergodicity of this flow and will use for this purpose a measure on the flow space which is derived from the Patterson/Sullivan measure developed in Chapters 3 and 4. The more classical theory is based upon Euclidean measure of the unit sphere and is to be regarded in some sense as a special case.

We start by defining the space on which the geodesic flow takes place. The space, denoted by Ω, comprises the set of all line elements in B. To be precise, at each point x of B we consider a direction — defined by a point ξ of S — and think of the pair (x,ξ) as representing a direction at x.

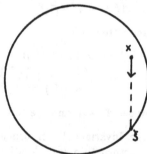

Figure 8.1.1

Thus Ω may be identified with $B \times S$. For $(x,\xi) \in \Omega$ we refer to x as the base point, or carrier point, of the line element. A Moebius transformation γ in $M(B)$ has a natural action on Ω defined by

$$\gamma(x,\xi) \;=\; \left[\gamma(x),\, \frac{\gamma'(x)}{|\gamma'(x)|}\,\xi\right].$$

We recall from section 7.2 that a flow acts on a metric space, and so we must find a suitable metric on Ω. Consider the function $d : \Omega \times \Omega \to R$ defined by

$$d((x,\xi),(y,\eta)) \;=\; \rho(x,y) + |\eta - \Delta(x,y)\,\xi|$$

where ρ is the hyperbolic metric in B and $\Delta(x,y)$ is the matrix defined in section 1.3. Not only is d a metric on Ω, but it is also invariant under the action of Moebius transformations.

Theorem 8.1.1 $\qquad$ The function d is a metric on Ω, and if $\gamma \in M(B)$ then $d(\gamma(x,\xi), \gamma(y,\eta)) = d((x,\xi),(y,\eta))$.

Proof. We start by proving the invariance. Since the hyperbolic metric in B is known to be invariant we have only to prove

$$\left| \frac{\gamma'(y)}{|\gamma'(y)|}\,\eta - \Delta(\gamma(x),\gamma(y))\,\frac{\gamma'(x)}{|\gamma'(x)|}\,\xi \right| \;=\; |\eta - \Delta(x,y)\,\xi|. \qquad (8.1.1)$$

However, from lemma 1.3.5, the left hand side of (8.1.1) is equal to

$$\left| \frac{\gamma'(y)}{|\gamma'(y)|}\,\eta - \frac{\gamma'(y)}{|\gamma'(y)|}\,\Delta(x,y)\,\xi \right|$$

and since $\gamma'(y)/|\gamma'(y)|$ is an orthogonal matrix the result follows immediately.

Trivially d is non-negative, and is equal to zero if and only if $x = y$ (for $\rho(x,y)$ is a metric) and $|\eta - \Delta(x,x)\,\xi| = 0$. But $\Delta(x,x)$ is the identity matrix, and so $d((x,\xi),(y,\eta)) = 0$ if and only if $(x,\xi) = (y,\eta)$.

In order to prove symmetry we must show

$$|\eta - \Delta(x,y)\,\xi| \;=\; |\xi - \Delta(y,x)\,\eta|. \qquad (8.1.2)$$

Multiplication by $\Delta(y,x)$ does not change length and so

$$|\eta - \Delta(x,y)\,\xi| \;=\; |\Delta(y,x)\,\eta - \Delta(y,x)\,\Delta(x,y)\,\xi|$$

but it is easy to check from the definition (1.3.7) that $\Delta(y,x)\,\Delta(x,y)$ is the identity and (8.1.2) follows.

Now for the triangle inequality. Given three points (x,ξ), (y,η), and (z,α) in Ω, we must show

$$d((x,\xi),(z,\alpha)) \;\leq\; d((x,\xi),(y,\eta)) + d((y,\eta),(z,\alpha)). \qquad (8.1.3)$$

Select a Moebius transformation γ in $M(B)$ with $\gamma(y) = 0$. Then, using a prime to denote the γ-image direction, we have

$$d((x,\xi),(y,\eta)) \;=\; d(\gamma(x,\xi),(0,\eta')) \;=\; \rho(\gamma(x),0) + |\eta' - \Delta(x,0)\xi'|$$

but $\Delta(x,0)$ is the identity and so $d((x,\xi),(y,\eta)) = \rho(\gamma(x),0) + |\eta' - \xi'|$. Similarly, $d((y,\eta),(z,\alpha)) = \rho(\gamma(z),0) + |\eta' - \alpha'|$. Thus the right hand side of (8.1.3) reduces to

$$\rho(\gamma(x),0) + \rho(\gamma(z),0) + |\eta' - \xi'| + |\eta' - \alpha'|$$

which is at least as big as $\rho(\gamma(x),\gamma(z)) + |\xi' - \alpha'|$ and, by the invariance of d, (8.1.3) is proved. $\square$

What does this metric really look like ? It is clear that the ρ component measures hyperbolic distance between base points. But what of the other part? If base points are the same ($x = y$) then $d((x,\xi),(x,\eta)) = |\xi - \eta|$ and we are basically measuring the angle between directions at x. Thus we think of the metric d as a sum of distance between base points and difference between angle at the base points.

Our next task is to put a measure on the space Ω. This can be done in several ways. To begin with, define a measure M on Ω by the differential

$$dM(x,\xi) \;=\; dV(x)\,dw(\xi)$$

where, as always, V is the hyperbolic volume and w is the $(n-1)$-dimensional Lebesgue measure on the unit sphere S. The measure M is obviously invariant under any Moebius transformation γ of $M(B)$ since V is invariant under γ, and the direction ξ undergoes a rotation which keeps the w-measure invariant.

Any line element (x,ξ) determines a directed geodesic which passes through x in the direction ξ. We write η_- and η_+ for the beginning and end points of this geodesic. Let z be the Euclidean mid-point of the geodesic and write s for the directed hyperbolic distance from z to x, the carrier point of our line element. Figure 8.1.2 illustrates the situation. We thus have a new set of coordinates on the space Ω and a natural correspondence

$$(x,\xi) \;\longleftrightarrow\; (\eta_-,\eta_+,s)$$

between the set of points (x,ξ) in Ω and the space $S \times S$ (minus diagonal) $\times R$. Using these new coordinates, we define a measure M' on Ω by means of the differential

$$dM' \;=\; \frac{2\,dw(\eta_-)\,dw(\eta_+)\,ds}{|\eta_+ - \eta_-|^{2n-2}}. \tag{8.1.4}$$

In fact this measure is nothing new.

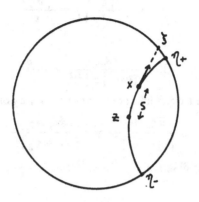

Figure 8.1.2

Theorem 8.1.1   The measures M, M' on Ω are equal.

Proof. To start with, we need to define the action of a Moebius transformation on Ω in terms of the coordinates (η_-,η_+,s). We set

$$\gamma(\eta_-,\eta_+,s) \;=\; (\gamma(\eta_-),\gamma(\eta_+),s+d)$$

where d is a real number defined as follows. If z is the Euclidean mid-point of the geodesic joining η_- to η_+ and w is the Euclidean mid-point of the geodesic joining $\gamma(\eta_-)$ to $\gamma(\eta_+)$ then d is the directed hyperbolic distance from w to $\gamma(z)$. Note that d depends on γ, η_-, and η_+ but is independent of s.

The measure M' is invariant under the action of any Moebius transformation. To see this, note that

$$dw(\gamma(\eta)) \;=\; |\gamma'(\eta)|^{n-1}\,dw(\eta)$$

and that, by (1.3.2)

$$|\gamma(\eta_+) - \gamma(\eta_-)| \;=\; |\gamma'(\eta_+)|^{1/2}\,|\gamma'(\eta_-)|^{1/2}\,|\eta_+ - \eta_-|.$$

Now, with M and M' both invariant under Moebius transformations, the Radon-Nikodym derivative is automorphic under the full Moebius group. But this group acts transitively on the space Ω and it follows that M' is a positive constant multiple of M.

It remains only to show that this constant is 1, and we consider first the two-dimensional situation. By $z = (\xi - i)/(\xi + i)$ we map the unit disc onto the upper half of the $\xi = u + iv$ plane, then

$$d\rho = \frac{2\,|\,dz\,|}{1 - |\,z\,|^2} = \frac{|\,d\xi\,|}{v}$$

and

$$dV = \frac{4\,dx\,dy}{(1 - |\,z\,|^2)^2} = \frac{du\,dv}{v^2}.$$

If η_1, η_2 are on the unit circle and are mapped to u_1, u_2 on the real axis then

$$\frac{d\eta_1 d\eta_2}{|\,\eta_1 - \eta_2\,|^2} = \frac{du_1 du_2}{|\,u_1 - u_2\,|^2}. \tag{8.1.5}$$

Consider the following diagram

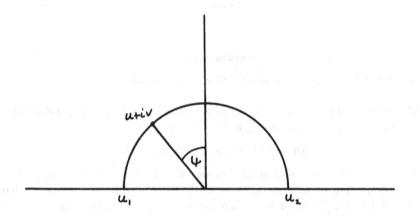

Figure 8.1.3

An easy calculation shows that the Jacobian of the transformation $(u,v) \to (u_1,u_2)$ is $\cos \psi/2$. Thus

$$dM = \frac{du\,dv\,d\psi}{v^2} = \frac{\cos \psi\, d\psi\, du_1\, du_2}{2v^2}.$$

But clearly,

$$v = \frac{|\,u_1 - u_2\,|}{2}\cos \psi \quad and \quad ds = \frac{|\,u_1 - u_2\,|}{2v}d\psi$$

and so

$$dM = \frac{2\,du_1\,du_2\,ds}{|\,u_1 - u_2\,|^2}$$

but this quantity is, by (8.1.5), dM' as required.

The proof proceeds by induction on the dimension n. It will thus be important for us to distinguish between several spaces and measures in the various dimensions. Write S^{n-1} for the unit sphere in R^n, B^n for the unit ball in R^n, w^{n-1} for the Lebesgue $(n-1)$-dimensional measure on S^{n-1}, and V^n for the hyperbolic volume in B^n. Then M^n and $(M')^n$ will denote the measures M and M' in dimension n. For the inductive step we proceed as follows. The ball B^{n-1} is viewed as the set $\{x \in B^n : x_n = 0\}$. Let D be a small ball in B^{n-1} centered at the origin, and consider $A = D \times S^{n-2}$, a subset of Ω^{n-1}. We assume that the theorem is true in dimension $n-1$ and so, in particular,

$$M^{n-1}(A) = V^{n-1}(D) \cdot w^{n-2}(S^{n-2}) = (M^{n-1})'(A) \qquad (8.1.6)$$

From A we now form a subset J of Ω^n as follows. First, let

$$D' = \{x = (x_1,...,x_n) : (x_1,...,x_{n-1}) \in D \text{ and } \rho(x,(x_1, \ldots, x_{n-1})) < \epsilon\}.$$

Now set

$$\Theta = \{\xi \in S^{n-1} : |\xi - S^{n-2}| < \epsilon\},$$

and write $J = D' \times \Theta$. Now

$$M^n(J) = V^n(D') \cdot w^{n-1}(\Theta)$$

$$\sim V^{n-1}(D) \cdot \epsilon \cdot w^{n-2}(S^{n-2}) \cdot \epsilon$$

$$= \epsilon^2 \cdot M^{n-1}(A) = \epsilon^2 \cdot (M^{n-1})'(A)$$

provided ϵ is small. We next need to compare $(M^{n-1})'(A)$ with $(M^n)'(J)$. Consider the cross-sectional diagram below. An easy calculation, using theorem 1.2.1, shows that $\delta \sim \epsilon$ for ϵ small, and thus the set of directions available at each point of D' is approximately 2ϵ. It is thus clear that

$$(M^n)'(J) \sim \frac{(2\epsilon)^2}{4} \cdot (M^{n-1})'(A)$$

and so, from (8.1.6), $(M^n)'(J) \sim M^n(J)$ for ϵ small. Knowing that $(M^n)'$ is a constant multiple of M^n, we deduce that this constant must be one, and the proof is complete. $\square$

We are now in a position to define a flow on the space Ω. To this end consider a function from $\Omega \times R$ onto Ω given by

$$[(\eta_-,\eta_+,s),t] \rightarrow (\eta_-,\eta_+,s+t).$$

This function forms the basis for the construction of the geodesic flow, and we write, for any real t,

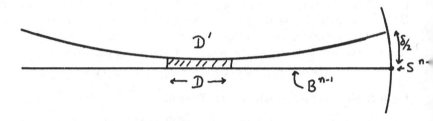

Figure 8.1.4

$$g_t(\eta_-,\eta_+,s) \;=\; (\eta_-,\eta_+,s+t).$$

Thus the action of g_t on a line element is simply to slide it a directed hyperbolic distance t along the geodesic it defines, preserving the direction of the geodesic. The reason for the term "geodesic flow" is now apparent.

It is trivial that the measure M' defined by (8.1.4) is invariant under the action of g_t and hence, by theorem 8.1.1, so is M. In fact, guided by (8.1.4), we may construct another measure on Ω, one which is derived from a conformal density. Let Γ be a discrete group preserving B, and let σ be a conformal density of dimension α. We further suppose that Γ is non-elementary and so has a positive exponent of convergence (corollary 3.4.5). It follows that α is positive (corollary 4.5.3) and, for $x \in B$, a measure m_x may be defined on Ω by

$$dm_x \;=\; \frac{d\,\sigma_x(\eta_-)\,d\,\sigma_x(\eta_+)\,ds}{|\eta_+ - \eta_-|^{2\alpha}} \qquad (8.1.7)$$

If $\gamma \in \Gamma$ and $\eta \in S$ then we recall, from the properties of a conformal density, that

$$\frac{d\,\sigma_x(\gamma(\eta))}{d\,\sigma_x(\eta)} \;=\; |\gamma'(\eta)|^{\alpha}$$

This, combined with (1.3.2) and the fact that

$$\gamma(\eta_-,\eta_+,s) \;=\; (\gamma(\eta_-),\gamma(\eta_+),s+d),$$

where d is independent of s, shows that m_x is invariant under $\gamma \in \Gamma$.

Given a discrete group Γ we move to consider the quotient space Ω/Γ. The map g_t has an action on Ω/Γ by virtue of the fact that it commutes with any Moebius transformation. In fact

$$\gamma(g_t(\eta_-,\eta_+,s)) \;=\; \gamma(\eta_-,\eta_+,s+t)$$

$$=\; (\gamma(\eta_-),\gamma(\eta_+),s+t+d)$$

$$=\; g_t(\gamma(\eta_-),\gamma(\eta_+),s+d)$$

$$=\; g_t(\gamma(\eta_-,\eta_+,s)) \;.$$

Invariance of the measures m_x, M under Moebius transformations leads to measures on the quotient space Ω/Γ which are also invariant under the map g_t. To obtain such a measure on Ω/Γ from M we proceed as follows (the m_x case is analogous). A subset A of Ω/Γ is said to be **measurable** if, with π denoting the natural projection map from Ω to Ω/Γ, $\pi^{-1}(A)$ is a Borel subset of Ω. If D_a denotes a Dirichlet region for Γ, define $A(D_a)$ as the set of points in $\pi^{-1}(A)$ with base points in D_a. We then define $M(A)$ to be the M measure of the set $A(D_a)$.

The metric d on Ω gives rise in a natural way to a metric on Ω/Γ (simply consider the minimal separation (d) of all lifts of a pair of points in Ω/Γ). With metric and measure defined on Ω/Γ it is routine to check that g_t is a flow on this space. It is called the geodesic flow. The terms g_t, M, m_x, and d will be used both on Ω and on Ω/Γ — no confusion should arise.

8.2 Basic Transitivity Properties

We begin by considering regional transitivity of the geodesic flow. With Γ a discrete group preserving B the flow g_t is regionally transitive when there is an element of Ω/Γ whose trajectory under the flow is dense in Ω/Γ. It follows immediately that the geodesic flow is regionally transitive if and only if there is a geodesic in B whose Γ-orbit is dense in the set of all geodesics. In other words, g_t is regionally transitive if and only if there is a line transitive point for Γ, and we have already seen that the line transitive set is non-empty precisely when Γ is of the first kind.

Theorem 8.2.1   The geodesic flow g_t on Ω/Γ is regionally transitive if and only if Γ is of the first kind.

The conservative and dissipative sets associated with a flow were defined in section 5.2. Consider the element (η_-,η_+,s) of Ω which we suppose to be a lift of a conservative point in Ω/Γ. As the geodesic flow is applied, we sweep out the geodesic ending at η_+ and return infinitely often to Γ-images of some fixed neighborhood of the origin. It follows that infinitely many members of $\Gamma(0)$ lie within a bounded distance of a half-geodesic ending at η_+. This makes η_+ a conical limit point. Thus a point of Ω/Γ belongs to the conservative set if and only if it determines a geodesic that ends at a conical limit point.

Recall that a flow is said to be conservative if the conservative set has full measure in the space. However, we have two measures — M and m_z — and will modify notation to say that the flow is conservative (M) or conservative (m_z). We next prove the following result.

Theorem 8.2.2   Let Γ be a discrete group preserving B with geodesic flow g_t defined on Ω/Γ.

(i) If the conical limit set has full w measure then g_t is conservative (M) otherwise the conical limit set has zero w measure and g_t is completely dissipative (M).

(ii) If the conical limit set has full σ_z measure then g_t is conservative (m_z) otherwise the conical limit set has zero σ_z measure and g_t is completely dissipative (m_z).

Proof. In view of the comments made just before the theorem it is clear that, in either case, the conservative set has full measure if and only if the conical limit set has full measure — with the appropriate measure on ∂B.

If the conical limit set has positive measure (w) then, by theorem 2.4.4, the group diverges at the exponent $n-1$ and is thus ergodic in its action on S, by corollary 5.2.4. Thus the conical limit set must in this case have full measure. This completes the proof of (i). Part (ii) of the theorem is an immediate consequence of theorem 4.4.4. $\square$

We conclude the section by showing that the conservative property of g_t is equivalent to divergence of the Poincaré series.

Theorem 8.2.3   Let Γ be a discrete group preserving B with geodesic flow g_t defined on Ω/Γ.

(i) The flow g_t is conservative (M) if and only if
$$\sum_{\gamma \in \Gamma} (1 - |\gamma(0)|)^{n-1} = \infty.$$

(ii) The flow g_t is conservative (m_z) if and only if
$$\sum_{\gamma \in \Gamma} (1 - |\gamma(0)|)^{\alpha} = \infty.$$

where α is the dimension of the conformal density σ giving rise to the measure m_z.

Proof. We have already done part (i), it is a combination of theorems 2.4.6 and 8.2.2. If the series given in part (ii) converges then by theorem 4.4.1 the σ_z measure of the conical limit set is zero and g_t is not conservative by theorem 8.2.2.

To show that the divergence of the series in (ii) implies that the flow g_t is conservative we will follow a method suggested by Sullivan [Sullivan, 1982 p.62]. This method can also be used to give another proof of theorem 2.4.6.

We note from the proof of theorem 4.3.2 that with $\epsilon > 0$ given, there exists R so that for $|a| > R$

$$\sigma_z(b(a:0,R)) \geq \epsilon .$$

Let Δ be the ball centered at the origin and of hyperbolic radius R. As in the proof of theorem 2.4.6, we may restrict attention to those $\gamma \in \Gamma$ such that $\gamma(\Delta) \cap \Delta = \varnothing$ without affecting the divergence of the series in (ii). Thus we assume from now on that $\gamma(\Delta) \cap \Delta = \varnothing$ for all $\gamma \in \Gamma - I$.

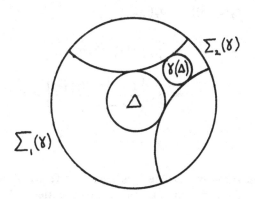

Figure 8.2.1

Now consider the situation in figure 8.2.1 where the product $\Sigma_1(\gamma) \times \Sigma_2(\gamma)$ determines the set of directed geodesics that pass first through Δ and then through $\gamma(\Delta)$. Note, from our remarks above, that

$$\sigma_z(\Sigma_1(\gamma)) \geq \epsilon \qquad\qquad (8.2.1)$$

for all $\gamma \in \Gamma$. Figure 8.2.2 illustrates the situation when γ^{-1} is applied to figure 8.2.1.

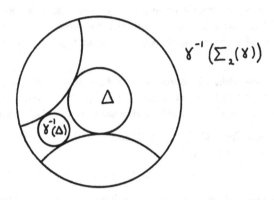

Figure 8.2.2

The set $\gamma^{-1}(\Sigma_2(\gamma))$ will be written Σ, and we note that $\sigma_x(\Sigma) \geq \epsilon$. Now

$$\sigma_x(\Sigma_2(\gamma)) \; = \; \sigma_x(\gamma(\Sigma))$$

$$= \; \int_\Sigma |\gamma'(\xi)|^\alpha d\sigma_x$$

$$= \; \int_\Sigma \left[\frac{1 - |\gamma^{-1}(0)|^2}{|\xi - \gamma^{-1}(0)|^2} \right]^\alpha d\sigma_x$$

$$> \; \lambda e^{-(0,\gamma 0)\alpha} \tag{8.2.2}$$

where λ is a group constant. Define J, a subset of Ω, by $J = \Delta \times S$, and note from the definition of the measure m_x, and the inequalities (8.2.1) and (8.2.2) that for any $\gamma \in \Gamma$

$$m_x(J \cap g_{-(0,\gamma 0)}\gamma(J)) \; > \; \lambda' \, e^{-(0,\gamma 0)\alpha}$$

where λ' is another group constant. If we assume that $R > 1$ and that $n \leq (0,\gamma 0) < n + 1$ then all line elements based in the shaded area of figure 8.2.3 and pointing towards $\gamma(\Delta)$ will be included in the set $J \cap g_{-n}\gamma(J)$ and it is easily seen that

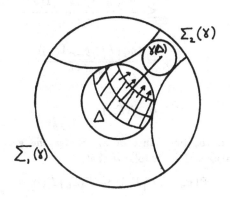

Figure 8.2.3

$$m_x(J \cap g_{-n}\gamma(J)) > k\, e^{-(0,\gamma 0)\alpha}$$

for a group constant k. Thus the divergence assumed in the theorem will ensure that

$$\sum_{n=1}^{\infty} m_x(J \cap g_{-n}\Gamma(J)) = \infty. \tag{8.2.3}$$

Now let E_n be the event $J \cap g_{-n}\Gamma(J)$ — i.e., the event that a point of J moves, under the geodesic flow, after time n into an image of J. The probability of this event is given by

$$P(E_n) = \frac{m_x(J \cap g_{-n}\Gamma(J))}{m_x(J)}.$$

We note that

$$P(E_{n+m} \cap E_n) = P(E_n) \cdot P(E_{n+m} \mid E_n),$$

and that there exists $\gamma \in \Gamma$ such that

$$P(E_{n+m} \mid E_n) = \frac{m_x(\gamma(J) \cap g_n(J) \cap g_{-m}\Gamma(J))}{m_x(\gamma(J))}$$

$$\leq \frac{m_x(\gamma(J) \cap g_{-m}\Gamma(J))}{m_x(J)}$$

$$= \frac{m_x(J \cap g_{-m}\Gamma(J))}{m_x(J)}$$

$$= P(E_m)$$

where we have used the invariance of m_x under the geodesic flow and under Moebius transformations of Γ. It follows that

$$P(E_{n+m} \cap E_n) \leq P(E_n) \cdot P(E_m).$$

We may now appeal to a Borel-Cantelli lemma (see [Billingsley, 1986 p.84] for example) to deduce that $P(E_n \text{ infinitely often}) = 1$. But this clearly means that almost every $x \in J$ is a conservative point for the flow g_t, and the proof of the theorem is complete. $\square$

8.3 Ergodicity

Our main aim in this section is to prove that the geodesic flow is ergodic if it is conservative. In the case of the measure M this has been proved by Hopf [Hopf, 1939] and in fact his ideas go through almost verbatim for the measure m_x. We will give the proof in full detail only for this latter case. We first need a lemma. Recall that d denotes the metric on the quotient space Ω/Γ.

Lemma 8.3.1   Let Γ be a discrete group preserving B and σ an α-dimensional conformal density invariant under Γ. There exists a positive, m_x integrable function λ on Ω/Γ such that for some real c and all x, $y \in \Omega/\Gamma$ with $d(x,y) \leq 1$

$$\frac{|\lambda(x) - \lambda(y)|}{\lambda(y)} < c\, d(x,y).$$

Proof. We consider the function $f\colon \Omega/\Gamma \to R^+$ defined as the distance of the carrier point of the line element to the orbit of the origin. Define, for $r > 0$, the set B_r as the f^{-1} image of the interval $[0,r]$. We first show that

$$\lim_{r \to \infty} \sup \frac{1}{r} \log m_x(B_r) < \infty. \tag{8.3.1}$$

If l is a geodesic passing through the ball of hyperbolic radius r centered at the origin then the hyperbolic length of the intersection of l with the ball is at most

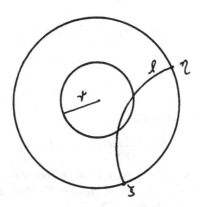

Figure 8.3.1

$2r$. If s is the hyperbolic distance from the origin to the geodesic then, by theorem 1.2.1,

$$\cosh s = \frac{2}{|\xi - \eta|}$$

and so, since $s \leq r$,

$$|\xi - \eta| > 4e^{-r}.$$

Thus the total m_z measure of all line elements lying on such geodesics is at most a constant times $re^{2\alpha r}$. But, on Ω/Γ, this set covers B_r and (8.3.1) is proved.

Choose $\epsilon > 0$ and define the function λ by

$$\lambda(x) = e^{-(2\alpha+\epsilon)f(x)}$$

noting that $\lambda > 0$ for all $x \in \Omega/\Gamma$. From the definition of f

$$\frac{|f(x) - f(y)|}{d(x,y)} \to 1 \quad as \quad d(x,y) \to 0$$

and, from this,

$$\frac{|\lambda(x) - \lambda(y)|}{\lambda(y)} = \frac{|e^{-(2\alpha+\epsilon)f(x)} - e^{-(2\alpha+\epsilon)f(y)}|}{e^{-(2\alpha+\epsilon)f(y)}}$$

$$= d(x,y)\frac{e^{-(2\alpha+\epsilon)f(z)}}{e^{-(2\alpha+\epsilon)f(y)}}$$

for some z between x and y. Now, for $d(x,y) < 1$, there is a constant k such that

$$e^{-[f(z)-f(y)]} \le e^{|f(z)-f(y)|}$$

$$< e^{|f(z)-f(y)|}$$

$$< e^{k \, d(z,y)} < e^k \,,$$

and we have a constant c such that

$$\frac{|\lambda(x) - \lambda(y)|}{\lambda(y)} < c \, d(x,y)$$

as required. It remains to show that λ is m_z integrable. On $B_{n+1} - B_n$, λ is at most $e^{-(2\alpha+\epsilon)n}$ and the m_z measure of $B_{n+1} - B_n$ does not exceed that of B_{n+1} which is at most a constant times $ne^{2\alpha n}$. Thus

$$\int_{B_{n+1}-B_n} \lambda \, dm_z < c \, ne^{-\epsilon n}$$

and so

$$\int_{\Omega/\Gamma} \lambda \, dm_z = \sum_{n=0}^{\infty} \int_{B_{n+1}-B_n} \lambda \, dm_z$$

$$< \sum_{n=0}^{\infty} C \, ne^{-\epsilon n} < \infty.$$

This completes the proof of the lemma.

The main result is the following.

Theorem 8.3.2   Let Γ be a discrete group preserving B and σ an α-dimensional conformal density invariant under Γ. Let m_z be the induced measure on Ω/Γ. If the geodesic flow is conservative (m_z) then it is ergodic (m_z).

Before embarking on the proof we note again that this theorem remains true if m_z is replaced by M. In this latter form it is due to Hopf [Hopf, 1939] — our present proof is modeled on his.

Proof. Let f, defined on Ω/Γ, be integrable (m_z) and let λ be the function constructed in lemma 8.3.1. Since g_t is conservative we have from the proof of theorem 7.2.10

$$\lim_{T\to\infty} \int_0^T \lambda(g_t(x))dt = +\infty$$

almost everywhere (m_z) on Ω/Γ. Now consider the quotient

$$F(T,x) = \frac{\int_0^T f(g_t(x))dt}{\int_0^T \lambda(g_t(x))dt}$$

which, by theorem 7.2.7, converges almost everywhere (as $T \to \infty$) to a measurable function that is invariant under g_t. We write

$$F(x) = \lim_{T \to \infty} F(T,x).$$

If we apply theorem 7.2.7 (with g replaced by λ, ϕ replaced by F, and h replaced by sign F) we obtain

$$\int_{\Omega/\Gamma} \lambda |F| \, dm_x \le \int_{\Omega/\Gamma} |f| \, dm_x . \qquad (8.3.2)$$

Suppose f_n is a sequence of L^1 functions converging (L^1) to f and we write

$$F_n(x) = \lim_{T \to \infty} \frac{\int_0^T f_n(g_t(x))dt}{\int_0^T \lambda(g_t(x))dt} .$$

Then from (8.3.2)

$$\int_{\Omega/\Gamma} \lambda |F_n - F| \, dm_x \to 0 \quad as \quad n \to \infty.$$

If each F_n is constant then, from some point on, these constants agree and F is constant almost everywhere. The point of these remarks is as follows. We wish to show that when f is integrable, F is constant almost everywhere — theorem 7.2.11 then states that g_t is ergodic. Our discussion above shows that it is sufficient to prove the constancy of F for continuous f with compact support, since such functions are dense in $L^1(m_x)$. We assume from now on that f is continuous with compact support.

The crucial step in the proof is to show that F takes the same value at points which determine asymptotic geodesics. In other words, suppose that $x, x' \in \Omega/\Gamma$ have the property that for a fixed real number a

$$d(g_t(x), g_{t+a}(x')) \to 0 \quad as \quad t \to +\infty.$$

Note that

$$F(x') = \lim_{T \to \infty} \frac{\int_0^T f(g_t(x'))dt}{\int_0^T \lambda(g_t(x'))dt}$$

$$= \lim_{T \to \infty} \frac{\int_0^T f(g_{t+a}(x'))dt}{\int_0^T \lambda(g_{t+a}(x'))dt}$$

where we have used the fact that $\int_0^T \lambda(g_t(x'))dt$ is unbounded. A straightforward calculation shows that the difference

$$\frac{\int_0^T f(g_t(x))dt}{\int_0^T \lambda(g_t(x))dt} - \frac{\int_0^T f(g_{t+a}(x'))dt}{\int_0^T \lambda(g_{t+a}(x'))dt}$$

is equal to the difference of

$$\frac{\int_0^T [(f(g_t(x))-f(g_{t+a}(x')))/\lambda(g_t(x))]\lambda(g_t(x))dt}{\int_0^T \lambda(g_t(x))dt} \qquad (8.3.3)$$

and

$$\frac{\int_0^T f(g_{t+a}(x'))dt}{\int_0^T \lambda(g_{t+a}(x'))dt} \cdot \frac{\int_0^T [(\lambda(g_t(x))-\lambda(g_{t+a}(x')))/\lambda(g_t(x))]\lambda(g_t(x))dt}{\int_0^T \lambda(g_t(x))dt} \cdot \qquad (8.3.4)$$

Since f is continuous with compact support and λ has the property given in lemma 8.3.1 we see that the expression given in (8.3.3) goes to zero as $T \to \infty$ and that the right hand factor in (8.3.4) does the same. The left factor in (8.3.4) remains bounded. Thus we have shown

$$F(x) = F(x').$$

The same conclusion is true if the elements x, x' are negative asymptotic in the sense that for some real b

$$d(g_{-t}(x), g_{-t+b}(x')) \to 0 \quad as \quad t \to +\infty.$$

The function F lifted to Ω determines a Γ-invariant function on $\partial B \times \partial B$ (minus diagonal) which, for almost all $\xi \in \partial B$, is constant on $\xi \times \partial B$ and on $\partial B \times \xi$. By Fubini's theorem, this function is constant almost everywhere on $\partial B \times \partial B$. We deduce that F is constant almost everywhere on Ω/Γ and the proof of the theorem is complete. $\square$

Ergodicity of the geodesic flow may be related to the group action on $\partial B \times \partial B$. Given a measure on ∂B (w or σ_x say) we have the product measure defined on $\partial B \times \partial B$ and the group action on this space defined in Chapter 6. These notions are connected to the geodesic flow via the following result.

Theorem 8.3.3   Let Γ be a discrete group preserving B.

(i) Γ is ergodic (w) on $\partial B \times \partial B$ if and only if g_t is ergodic (M).

(ii) Γ is ergodic (σ_x) on $\partial B \times \partial B$ if and only if g_t is ergodic (m_x).

Proof.   We prove (i) as the proof of (ii) is identical. If Γ is not ergodic on

$\partial B \times \partial B$ then there exists a subset Q of $\partial B \times \partial B$ which is invariant under Γ and with the property that both Q and its complement have positive measure $(w \times w)$. Define a subset Q' of Ω by

$$(\xi, \eta, s) \in Q' \iff (\xi, \eta) \in Q.$$

The set Q' is clearly of positive measure (M) in Ω and, being invariant under Γ, gives rise to a subset of positive measure in Ω/Γ. This subset is clearly g_t-invariant and the same comments apply to its complement. It follows that g_t is not ergodic. The argument is clearly reversible and the proof of the theorem is complete . $\square$

To conclude this section we relate ergodicity to group divergence at various exponents. The results below summarize what is known concerning ergodic phenomena relating to the two measures M and m_x.

Theorem 8.3.4   Let Γ be a discrete group preserving B with measure M on the line element space Ω/Γ derived from Lebesgue $(n-1)$-dimensional measure w on ∂B. The following are equivalent:

1.   The conical limit set has full measure (w).

2. g_t is conservative (M).

3. g_t is ergodic (M).

4. Γ is ergodic on $\partial B \times \partial B$ $(w \times w)$.

5. $\sum\limits_{\gamma \in \Gamma} (1 - |\gamma(0)|)^{n-1} = \infty$.

Proof.   Theorem 8.2.1 shows $1 \iff 2$, theorem 8.3.2 shows $2 \iff 3$, and theorem 8.3.3 shows $3 \iff 4$. The equivalence of (1) and (5) is given in theorem 6.3.3. $\square$

In terms of measures derived from a conformal density we have the following.

Theorem 8.3.5   Let Γ be a discrete group preserving B and σ a Γ-invariant conformal density of dimension α. The following are equivalent:

1.   The conical limit set has full measure (σ_x).

2. g_t is conservative (m_x).

3. g_t is ergodic (m_x).

4. Γ is ergodic on $\partial B \times \partial B$ $(\sigma_x \times \sigma_x)$.

5. $\displaystyle\sum_{\gamma \in \Gamma} (1 - |\gamma(0)|)^\alpha = \infty.$

Proof. The equivalence of the first four properties has, as we remarked in the proof of theorem 8.3.4, been established in this chapter. Theorem 8.2.3 shows that (1) is equivalent to (5). □

In terms of a manageable criterion to check for the ergodicity of the geodesic flow, we should recall theorem 7.2.9 which shows that if the space Ω/Γ has finite total measure then the flow g_t is conservative — hence ergodic. The following result is an easy consequence of the definition of the measure M on the quotient space.

Theorem 8.3.6   Let Γ be a discrete group preserving B and with Dirichlet region D_0 centered at the origin, then $M(\Omega/\Gamma) < \infty$ if and only if the hyperbolic volume of D_0 is finite.

In the case of the measures m_z it is very difficult to check whether $m_z(\Omega/\Gamma) < \infty$. In fact we will be devoting the next chapter to this topic. However, as the next result shows, this is a worthy endeavor.

Theorem 8.3.7   If Γ is a non-elementary discrete group with a conformal density σ such that $m_z(\Omega/\Gamma) < \infty$ then

1.   the dimension of σ is $\delta(\Gamma)$ — the exponent of convergence of Γ.

2. σ is unique.

3. σ gives full measure to the conical limit set.

4. g_t is ergodic (m_z).

Proof. Parts (1) and (2) follow from (3) as in Chapter 4, and (4) follows from (3) as we have seen in theorem 8.3.5. Property (3) is true by theorems 7.2.9 and 8.2.2. □

CHAPTER 9

Geometrically Finite Groups

9.1 Introduction

If Γ is a discrete group preserving B and σ is a Γ-invariant conformal density of dimension α then we have seen in the previous chapter how to define from σ a measure m_x on the quotient line element space Ω/Γ. For our major applications we will work with groups Γ for which $m_x(\Omega/\Gamma)$ is finite. Our aim in the first two sections of this chapter is to show that geometrically finite groups have this property. In this first section we develop a test due to Sullivan [Sullivan, 1979] which gives a sufficient condition for $m_x(\Omega/\Gamma)$ to be finite.

We start by considering the average time which a trajectory spends in a bounded set Q of Ω/Γ. The m_x measure of a bounded set is finite, and the flow g_t is measure preserving. If the flow is dissipative then almost all trajectories spend only a finite amount of time in Q and we have

$$\lim_{T \to \infty} \frac{1}{T} \int_0^T 1_Q(S_t(v)) dt = 0$$

almost everywhere. If, on the other hand, the flow is conservative then it is ergodic, and we know from the ergodic theorem that

$$\lim_{T \to \infty} \frac{1}{T} \int_0^T 1_Q(g_t(v)) dt = k$$

where $k = 0$ unless $m_x(\Omega/\Gamma) < \infty$ in which case

$$k = \frac{m_x(Q)}{m_x(\Omega/\Gamma)} .$$

Putting these ideas together we have the following lemma.

Lemma 9.1.1   Let Γ be a discrete group and Q a bounded subset of Ω/Γ. There exists a constant k such that for almost all (m_x) v in Ω/Γ

$$\lim_{T\to\infty} \frac{1}{T}\int_0^T 1_Q(g_t(v))dt = k.$$

The value of k is zero unless $m_x(\Omega/\Gamma) < \infty$, and in this case

$$k = \frac{m_x(Q)}{m_x(\Omega/\Gamma)}.$$

Using this lemma we will next prove a condition which is necessary for $m_x(\Omega/\Gamma) = \infty$.

Theorem 9.1.2   Let Γ be a discrete group with a conformal density σ of dimension α. If $m_x(\Omega/\Gamma) = \infty$, where m_x is the measure derived from the conformal density, then

$$\lim_{T\to\infty} \frac{1}{T}\sum_{\substack{\gamma\in\Gamma \\ (x,\gamma x)\leq T}} e^{-\alpha(x,\gamma x)} = 0.$$

Proof. We will consider the case $x = 0$, for the general situation is an easy modification of this. We denote a generic point of Ω by $v = (\xi,\eta,s)$. Now if $Q \subset \Omega$ and $\xi \in S$ we define

$$Q_\xi = (\xi \times S \times R) \cap Q$$

in other words, Q_ξ represents the points of Q lying on geodesics starting at ξ.

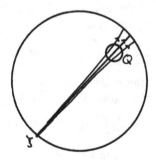

Figure 9.1.1

On Q_ξ we consider the measure

$$d\lambda = \frac{d\sigma_x(\eta)ds}{|\xi - \eta|^{2\alpha}}$$

and note that

$$dm_x = d\sigma_x(\xi)\, d\lambda\,.$$

Choose a ball Δ centered at the origin, and let the set Q be the collection of all line elements based at points of Δ. If $\xi \in \partial B$ and $T > 0$ we define

$$F(T,\xi) = \frac{1}{T}\, \lambda\{g_t(Q_\xi) \cap \bigcup_{\substack{\gamma \in \Gamma \\ (0,\gamma 0) < T}} \gamma(Q)\}/\lambda(Q_\xi)\,.$$

This quantity is the average time spent by the flow lines starting in Q_ξ in images of Q to a distance T from the origin. It is evident that if r denotes the hyperbolic radius of Δ

$$F(T,\xi) \leq \frac{1}{\lambda(Q_\xi)} \int Q_\xi \left[\frac{1}{T}\int_0^{T+2r} 1_{\Gamma(Q)}\, g_t(v)dt\right] d\lambda(v)\,.$$

By lemma 9.1.1, the integrand above converges to 0 — provided that v avoids a certain set of measure 0. So by the bounded convergence theorem

$$\lim_{T \to \infty} F(T,\xi) = 0. \qquad (9.1.1)$$

An alternative way to write $F(T,\xi)$ is as follows

$$F(T,\xi) = \frac{1}{T}\sum_{\substack{\gamma \in \Gamma \\ (0,\gamma 0) < T}} \frac{\lambda(g_{-t}(\gamma(Q)) \cap Q_\xi)}{\lambda(Q_\xi)}$$

$$= \frac{1}{T\,\lambda(Q_\xi)}\sum_{\substack{\gamma \in \Gamma \\ (0,\gamma 0) < T}} \int Q_\xi 1_{g_{-t}(\gamma(Q))}(v)\, \frac{d\sigma(\eta)ds}{|\xi - \eta|^{2\alpha}}\,.$$

If, in the sum appearing in the last expression, we were to confine attention to those $\gamma \in \Gamma$ such that $\gamma(0)$ is close (say within Euclidean distance ϵ) to the point ξ' antipodal to ξ, then we could assert that the factor $|\xi - \eta|^{2\alpha}$ is bounded above and below, and hence also the arc length contribution to the measure. Denoting by $b(y\!:\!x,c)$ the projection of a ball centered at x and of radius c from y onto ∂B (this is the notation of section 4.3) we have, for $\epsilon > 0$ chosen and a constant k depending on this choice,

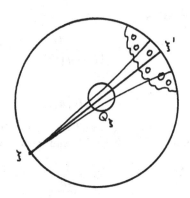

Figure 9.1.2

$$F(T,\xi) \geq \frac{1}{T\,\lambda(Q_\xi)} \sum_{\substack{\gamma \in \Gamma \\ (0,\gamma 0) < T, |\overline{\gamma(0)} - \xi'| < \epsilon}} 1_{g_{-1}\gamma(Q)}(v) \frac{d\,\sigma(\eta)ds}{|\xi - \eta|^{2\alpha}}$$

$$> \frac{K}{T\,\lambda(Q_\xi)} \sum_{\substack{\gamma \in \Gamma \\ (0,\gamma 0) < T, |\overline{\gamma(0)} - \xi'| < \epsilon}} \sigma(b(\xi\colon \gamma(0), r)) \,.$$

By (9.1.1) we obtain

$$\lim_{T \to \infty} \sum_{\substack{\gamma \in \Gamma \\ (0,\gamma 0) < T, |\overline{\gamma(0)} - \xi'| < \epsilon}} \sigma(b(\xi\colon \gamma(0), r)) = 0 \,. \tag{9.1.2}$$

If $\epsilon > 0$ is small then, since $|\gamma(0) - \xi'| < \epsilon$, $b(\xi, \gamma(0), r)$ is "close" to $b(0, \gamma(0), r)$. To be more precise, we choose $0 < R < R'$ so large that if $r \in (R, R')$ then

$$b(0\colon \gamma(0), R) \subset b(\xi\colon \gamma(0), r) \subset b(0\colon \gamma(0), R')$$

and also to ensure that we may apply theorem 4.3.2 to $b(0\colon \gamma(0), R)$. Thus, by that theorem, we will have from (9.1.2) above

$$\frac{1}{T} \sum_{\substack{\gamma \in \Gamma \\ (0,\gamma 0) < T, |\overline{\gamma(0)} - \xi'| < \epsilon}} r(0\colon \gamma(0), R)^\alpha \to 0 \quad as \quad T \to \infty.$$

Now the limit set of Γ is covered by a finite number of balls of radius ϵ centered at points ξ' antipodal to points ξ as above. Thus we may deduce from the above that

$$\lim_{T \to \infty} \frac{1}{T} \sum_{\substack{\gamma \in \Gamma \\ (0,\gamma 0) < T}} r\,(0 : \gamma(0), R\,)^\alpha = 0.$$

We now use the formula (4.3.1) for $r\,(0 : \gamma(0), R\,)$ and see that

$$\lim_{T \to \infty} \frac{1}{T} \sum_{\substack{\gamma \in \Gamma \\ (0,\gamma 0) < T}} (1 - |\gamma(0)\,|)^\alpha = 0$$

or equivalently,

$$\lim_{T \to \infty} \frac{1}{T} \sum_{\substack{\gamma \in \Gamma \\ (0,\gamma 0) < T}} e^{-\alpha(0,\gamma 0)} = 0$$

as required. □

Using this result we will derive Sullivan's criterion for a group to have a quotient line element space of finite volume.

Theorem 9.1.3   Let Γ be a non-elementary discrete group with critical exponent δ. If the Poincaré series satisfies

$$\sum_{\gamma \in \Gamma} e^{-\delta(z,\gamma z)} \geq \frac{A}{s - \delta} \text{ for } s > \delta$$

where A is a constant, then $m_z (\Omega/\Gamma) < +\infty$.

Proof. Since Γ is non-elementary, we know that $\delta > 0$ and that a conformal density σ of dimension δ exists. For positive integer k write n_k for the cardinality of the set

$$\{\gamma \in \Gamma : k - \frac{1}{2} \leq (0,\gamma 0) < k + \frac{1}{2}\}$$

and note from theorem 4.5.1 that there is a constant C with

$$n_k \leq C e^{k\delta}.$$

The Poincaré series $\sum_{\gamma \in \Gamma} e^{-s(0,\gamma 0)}$ with $s > \delta$ is proportional to the series $\sum_{k=1}^{\infty} n_k e^{-sk}$ which is majorized by the series $\sum_{k=1}^{\infty} C e^{-(s-\delta)k}$ and so for a (different) constant C we will have

$$\sum_{\gamma \in \Gamma} e^{-s(0,\gamma 0)} < \frac{C}{s - \delta}. \tag{9.1.3}$$

Considering the tail of the series,

$$\sum_{k=T+1}^{\infty} n_k e^{-sk} < \sum_{k=T+1}^{\infty} B e^{k(\delta-s)}$$

$$= B e^{(T+1)(\delta-s)} \sum_{k=0}^{\infty} e^{k(\delta-s)}$$

$$< \frac{D e^{T(\delta-s)}}{s-\delta} \qquad\qquad (9.1.4)$$

for some constant D. We suppose now that for some constant F,

$$\sum_{\gamma \in \Gamma} e^{-s(0,\gamma 0)} \geq \frac{F}{s-\delta}$$

for $s > \delta$, and choose τ so large that if $T \geq \tau$ then

$$\frac{D}{T e^{T(s-\delta)}} < \frac{F}{2}.$$

If follows then from (9.1.4) that

$$\frac{1}{T} \sum_{\substack{\gamma \in \Gamma \\ (0,\gamma 0)<T}} e^{-s(0,\gamma 0)} > \frac{F}{(s-\delta)T} - \frac{D e^{T(\delta-s)}}{T(s-\delta)}$$

and, provided $T \geq \tau$, the quantity on the right exceeds

$$\frac{F}{2(s-\delta)T}.$$

Now consider a sequence $\{s_n\}$ monotonic decreasing to δ and define $T_n = 1/(s_n - \delta)$ then we will have

$$\frac{1}{T_n} \sum_{(0,\gamma 0)<T_n} e^{-s_n(0,\gamma 0)} > \frac{F}{2}$$

however, since $s_n > \delta$,

$$\frac{1}{T_n} \sum_{(0,\gamma 0)<T_n} e^{-\delta(0,\gamma 0)} > \frac{F}{2}$$

and, using theorem 9.1.2, the proof is complete. □

9.2 Volume of the Line Element Space

Patterson has shown [Patterson, 1976a p.266] that finitely generated Fuchsian groups satisfy the hypothesis of theorem 9.1.3 and so for such groups $m_x(\Omega/\Gamma) < \infty$. In this section we shall be proving a result of Sullivan which generalizes the result to any geometrically finite discrete group. We start with a simple geometric lemma. The proof is a routine calculation and will be omitted.

Lemma 9.2.1   Suppose σ is a geodesic segment of hyperbolic length T and, for $k \geq 0$, let A_k be the set of points in B within a hyperbolic distance k of σ. Then there exists a constant C, depending only on k, such that

$$V(A_k) = CT.$$

where V is the hyperbolic volume.

From now on, for the remainder of the section, we assume that Γ is a geometrically finite discrete group acting on B. We recall some notation defined in section 5.3. The Dirichlet region for Γ centered at the origin will be denoted D_0 and we intersect D_0 with the set of points which are at most a hyperbolic distance of K from the convex hull of the limit set of Γ. This intersection will be written $D(K)$. For $\xi, \eta \in S$ and $K > 0$ write $D_{\xi,\eta}(K)$ for the intersection of $D(0)$ with the set of points distant at most K from the geodesic joining ξ and η. It is clear that

$$D(K) = \bigcup D_{\xi,\eta}(K)$$

where the union is taken over those pairs (ξ,η) with the property that the geodesic joining ξ to η lies in the convex hull of the limit set.

With all this notation in place we proceed as follows. Consider the function ϕ_μ defined by (5.3.2) then,

$$\int_{D(K)}[\phi_\mu(x)]^2 dV = 4^\delta \int_{D(K)} \int_{S \times S} \frac{1}{|\xi - \eta|^{2\delta}}[\cosh s_{\xi\eta}(x)]^{-2\delta} d\mu_0(\xi)d\mu_0(\eta)dV$$

from lemma 5.3.2, where $s_{\xi,\eta}(x)$ denotes the hyperbolic distance from the point x to the geodesic joining ξ to η. Interchanging the order of integration we obtain

$$\int_{D(K)}[\phi_\mu(x)]^2 dV = 4^\delta \int_{S \times S} \frac{1}{|\xi - \eta|^{2\delta}}\left[\int_{D_{\xi\eta}(K)} \frac{1}{[\cosh s_{\xi\eta}(x)]^{2\delta}} dV\right] d\mu_0(\xi)d\mu_0(\eta).$$

$$> \frac{4^\delta}{[\cosh(K)]^{2\delta}} \int_{S \times S} \frac{1}{|\xi - \eta|^{2\delta}} V(D_{\xi\eta}(K))d\mu_0(\xi)d\mu_0(\eta).$$

Now we use lemma 9.2.1 which tells us that the volume term $V(D_{\xi\eta}(K))$ in the above integral is bounded below by a constant (depending only on K) times the hyperbolic length of the geodesic $D_{\xi\eta}(0)$. Thus, for a constant C depending only upon Γ and K, we have

$$\int_{D(K)}[\phi_\mu(x)]^2 dV > C \int_{S \times S} \frac{1}{|\xi - \eta|^{2\delta}} \int_{D_{\xi\eta}(0)} dt \; d\mu_0(\xi)d\mu_0(\eta).$$

The right hand side above is the m_μ measure of a set of line elements — namely those line elements in D_0 which determine geodesics lying in the convex hull of the limit set. Since the measure μ_0 is concentrated on the limit set we might just as well extend this set of line elements to include all of those based at points of D_0. Then the inequality above becomes

$$\int_{D(K)} [\phi_\mu(x)]^2 dV > C \int_{\Omega/\Gamma} dm_\mu = Cm_\mu(\Omega/\Gamma). \qquad (9.2.1)$$

This condition for a geometrically finite discrete group to have a quotient line element space of finite (m_μ) volume is just what we need. Combining equation (9.2.1) with theorem 5.3.3, we have proved the main theorem.

Theorem 9.2.2   [Sullivan 1984] If Γ is a geometrically finite discrete group and m_μ is the measure on Ω/Γ derived from the canonical conformal density of dimension δ then

$$m_\mu(\Omega/\Gamma) < \infty.$$

9.3 Hausdorff Dimension of the Limit Set

The reader will recall from chapter 4 that, for a convex co-compact group, the Hausdorff dimension of the limit set is equal to the critical exponent of the group. We start this section by extending this result to groups Γ for which $m_x(\Omega/\Gamma) < \infty$. In particular, the result is true for geometrically finite groups. We write δ $(=\delta(\Gamma))$ for the critical exponent of Γ and μ for the conformal density of dimension δ whose existence is assured by theorem 4.1.2. We suppose throughout this section that $m_x(\Omega/\Gamma) < \infty$, where m_x is the line element measure derived from μ — note that this property is guaranteed for any geometrically finite group. From our discussions in the last section we know also that Γ diverges at the critical exponent.

We start with a result whose proof is an immediate (and minor) modification of the proof of theorem 2.4.4.

Theorem 9.3.1   Let Γ be a discrete group acting in B. If, for some $\beta > 0$,

$$\sum_{\gamma \in \Gamma} (1 - |\gamma(0)|)^\beta < \infty$$

then the conical limit set has zero β-dimensional Hausdorff measure.

As a consequence we have

Corollary 9.3.2   If Γ is a discrete group with critical exponent δ then the Hausdorff dimension of the conical limit set is at most δ.

Thus, for any discrete group, $d(C) \leq \delta(\Gamma)$. We now consider the inequality in the opposite direction. We need a lemma.

Lemma 9.3.3   If $m_x(\Omega/\Gamma) < \infty$ and if $\sigma(t,v,u)$ denotes the distance from $g_t(v)$ to u then, with u fixed, for almost all (m_x) $v \in \Omega/\Gamma$,

$$\lim_{t \to \infty} \frac{1}{t} \sigma(t,v,u) = 0.$$

Proof.   Define the function $\psi(v)$ on Ω/Γ to be the directional derivative of $\sigma(t,v,u)$ in the direction v. Observe that $|\psi(v)| \leq 1$ for all $v \in \Omega/\Gamma$ and so, since $m_x(\Omega/\Gamma) < \infty$, ψ is integrable (dm_x) over Ω/Γ. Now since ψ takes opposite values at line elements with the same base point, but pointing in the opposite direction, it is immediate that

$$\int_{\Omega/\Gamma} \psi(v)dm_x = 0.$$

Now g_t is ergodic (since $m_x(\Omega/\Gamma) < \infty$) and by the ergodic theorem, for almost all (m_x) $v \in \Omega/\Gamma$,

$$\lim_{T \to \infty} \frac{1}{T}\int_0^T \psi(g_t(v))dt = \int_{\Omega/\Gamma} \psi(v)dm_x = 0.$$

But

$$\frac{1}{T}\int_0^T \psi(g_t(v))dt = \frac{1}{T}[\sigma(T,v,u) - \sigma(0,v,u)]$$

and the lemma is proved. □

   The next lemma is the heart of the matter — balls centered over a large part of the conical limit set, have μ_x measure not exceeding a constant times $r^{\delta-\epsilon}$. This, as we shall see, is exactly the property needed for the lower bound on the Hausdorff dimension of the conical limit set.

Lemma 9.3.4   Let Γ be a discrete group with critical exponent δ and satisfying $m_x(\Omega/\Gamma) < \infty$ for the measure m_x derived from the canonical conformal density μ. If C denotes the conical limit set then there exists a compact subset K of C such that $\mu_x(K) > 0$ and for any $\epsilon > 0$ there exists $r_0 > 0$ such that if $\xi \in K$ and $r < r_0$

$$\mu_x(B(\xi,r))/r^{\delta-\epsilon} < A$$

were $B(\xi,r)$ is the ball in S center ξ radius r intersected with K.

Proof.   Given $t > 0$ and $\xi \in S$ we define a region $D_{t,\xi} \subset \overline{B}$ in the following way. Let x_t be the point on the radius to ξ which is distant t from the origin and then we say that $u \in D_{t,\xi}$ if the geodesic joining u to x_t makes an angle $> \frac{\pi}{2}$ with the geodesic joining 0 to x_t. From the hyperbolic cosine rule [Beardon, 1983 p.148] we have that $D_{t,\xi}$ is the closure of the set

$$\{u \in B : \cosh(0,u) > \cosh(x_t,u)\cdot\cosh t\}.$$

If $\gamma(0) \in D_{t,\xi}$ then

$$e^{(0,\gamma 0)} > \cosh(0,\gamma 0) > \cosh(x_t,\gamma 0)\cdot\cosh t$$

$$> \frac{1}{4}e^{(x_t,\gamma 0)}e^{\,t}$$

and so if $s > \delta$ is chosen

$$\sum_{\gamma\,:\,\gamma 0\,\in\,D_{t,\xi}} e^{-s(0,\gamma 0)} < 4^s\, e^{-st} \sum_{\gamma\,:\,\gamma 0\in D_{t,\xi}} e^{-s(x_t,\gamma 0)}.$$

If $\gamma_0(0)$ is the closest image of 0 to x_t then, writing $\sigma(t) = (x_t,\gamma_0(0))$, we have from the triangle inequality

$$(x_t,\gamma(0)) \ge (\gamma_0(0),\gamma 0) - \sigma(t)$$

from which

$$e^{st}\sum_{\gamma\,:\,\gamma 0\,\in\,D_{t,\xi}} e^{-s(0,\gamma 0)} < 4^s\, e^{s\sigma(t)} \sum_{\gamma\,:\,\gamma 0\,\in\,D_{t,\xi}} e^{-s(\gamma_0(0),\gamma 0)}$$

$$< 4^s\, e^{s\sigma(t)} \sum_{\gamma\in\Gamma} e^{-s(\gamma_0(0),\gamma 0)}$$

$$= 4^s\, e^{s\sigma(t)} \sum_{\gamma\in\Gamma} e^{-s(0,\gamma 0)}.$$

Thus, using the notation of section 3.1 (noting that the group Γ diverges at its critical exponent), we have

$$\mu_{0,s}(D_{t,\xi}) < \frac{4^s\, e^{s\sigma(t)}}{e^{st}}.$$

If we write $B_{t,\xi}$ for $D_{t,\xi} \cap S$ then, from Helly's theorem and the definition of the measure μ_0, we obtain from the above

$$\mu_0(B_{t,\xi}) \le \frac{4^\delta e^{\delta\sigma(t)}}{e^{\delta t}}. \qquad (9.3.1)$$

Now $B_{t,\xi}$ is a ball in S centered at ξ and of radius r say. We next need to compute r in terms of t. The region $D_{t,\xi}$ is the intersection of $\overline{B}$ with an open Euclidean ball centered at ω say and of Euclidean radius λ where [Beardon, 1983 p.157]

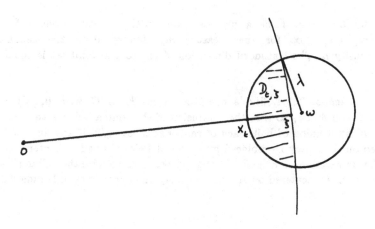

Figure 9.3.1

$$\lambda = \frac{1}{\sinh t}.$$

Thus $r = \arctan((\sinh t)^{-1})$ and so, since $4e^{-t} > (\sinh t)^{-1} > 2e^{-t}$, and $\tan\theta/\theta \to 1$ as $\theta \to 0$, we see that for $t > t_0$ say

$$8e^{-t} > r > e^{-t}.$$

With r as above and $\epsilon > 0$ chosen,

$$\frac{\mu_0(B_{t,\xi})}{r^{\delta-\epsilon}} \leq \frac{4^\delta e^{\delta\sigma(t)}}{e^{\delta t} e^{-t(\delta-\epsilon)}}$$

$$= 4^\delta e^{\delta\sigma(t)-\epsilon t}$$

provided $t > t_0$. It is a consequence of lemma 9.3.3 that, for almost all ξ, $\frac{\sigma(t)}{t} \to 0$ as $t \to \infty$. We may find a compact subset K of C which has positive measure and with the further property that if $t > t_1$ and $\xi \in K$ then $\frac{\sigma(t)}{t} < \frac{\epsilon}{\delta}$. For such values of t

$$\frac{\mu_0(B_{t,\xi})}{r^{\delta-\epsilon}} \leq 4^\delta$$

and the lemma is proved. $\square$

The main result is as follows.

Theorem 9.3.5   Let Γ be a discrete group with critical exponent δ and satisfying $m_z(\Omega/\Gamma) < \infty$ for the measure m_z derived from the canonical conformal density μ. The Hausdorff dimension of the conical limit set is equal to δ.

Proof. As in lemma 9.3.4, find a compact subset K of C with $\mu_z(K) > 0$. Choose $\epsilon > 0$ and find r_0 so that the conclusion of the lemma holds with $r < r_0$. Now cover K by a union of balls each of radius at most $r_0/2$. If any such ball is not centered on K it may be replaced by a ball of twice the radius centered on a point in the intersection of K and the original ball. We assume then that all the balls in the cover are centered on K and so satisfy the hypotheses of lemma 9.3.4. We write

$$K \subset \bigcup_i B_i$$

let r_i be the radius of B_i and note that

$$\sum_i r_i^{\delta-\epsilon} \geq A \sum_i \mu_z(B_i)$$

$$\geq A \, \mu_z(\bigcup_i B_i)$$

$$\geq A \, \mu_z(K) > 0 \, .$$

Thus K has positive $(\delta-\epsilon)$-dimensional Hausdorff measure. We observe then that the Hausdorff dimension of K is at least $\delta - \epsilon$. Thus

$$d(C) \geq \delta - \epsilon$$

for every $\epsilon > 0$ and so $d(C) \geq \delta$. Combining this with corollary 9.3.2, the proof of the theorem is complete. $\square$

For a geometrically finite group the limit set comprises the conical limit set and a countable collection of parabolic cusps and we have the next result.

Theorem 9.3.6   If Γ is a geometrically finite discrete group then the Hausdorff dimension of the limit set is equal to the critical exponent.

We can use our ideas to prove a recent deep theorem of Sullivan [Sullivan, 1984] and Tukia [Tukia, 1984].

Theorem 9.3.7   If Γ is a geometrically finite discrete group of the second kind acting in the unit ball B of R^n and if Λ denotes the limit set of Γ then

$$d(\Lambda) < n - 1.$$

Proof. Let δ denote the critical exponent of the group and note that

$$\sum_{\gamma \in \Gamma} (1 - |\gamma(0)|)^{\delta} = \infty$$

but theorem 1.6.2 tells us that

$$\sum_{\gamma \in \Gamma} (1 - |\gamma(0)|)^{n-1} < \infty$$

and so $\delta < n - 1$. We now apply theorem 9.3.6 to complete the proof. $\square$

If we specialize to dimension $n = 2$ (Fuchsian groups) we obtain better results concerning the Hausdorff dimension of the limit set.

Theorem 9.3.8 If Γ is a finitely generated Fuchsian group then the Hausdorff dimension of the limit set is $\delta(\Gamma)$.

Proof. This is an immediate consequence of the fact that in the two-dimensional case geometrically finite means finitely generated. $\square$

Theorem 9.3.9 For an arbitrary Fuchsian group the Hausdorff dimension of the conical limit set is $\delta(\Gamma)$.

Proof. Any Fuchsian group Γ can be written as a union of finitely generated groups

$$\Gamma = \bigcup_{\alpha} \Gamma_{\alpha}.$$

If we write C_{α} for the conical limit set of Γ_{α}, and C for the conical limit set of Γ then clearly

$$C_{\alpha} \subset C$$

and so

$$d(C) \geq \sup_{\alpha} d(C_{\alpha}) = \sup_{\alpha} \delta_{\alpha} \qquad (9.3.2)$$

where δ_{α} is the critical exponent of Γ_{α}. Now choose invariant conformal densities $\mu(\alpha)$ for Γ_{α} of dimension δ_{α}. Find a sequence of indices so that $\lim_{i \to \infty} \delta_i = \sup_{\alpha} \delta_{\alpha} = \delta'$ say, and $\mu_x(i)$ converges weakly to μ_x' (as $i \to \infty$). Any element of Γ is eventually in Γ_{α} and $\gamma^* \mu_x(i) = \dfrac{1}{|\gamma'|}\delta_i \mu_x(i)$ since $\mu(i)$ is an invariant conformal density. Letting $i \to \infty$ yields

$$\gamma^* \mu_z{}' = \frac{1}{|\gamma'|} \delta' \mu_z{}' \ .$$

But this is true for each $\gamma \in \Gamma$, and so $\mu_z{}'$ determines an invariant conformal density of dimension δ'. It follows from theorem 4.5.3 that $\delta' \geq \delta$ — the critical exponent of Γ. From (9.3.2) it follows that $d(C) \geq \delta$, and corollary 9.3.2 completes the proof. $\square$

CHAPTER 10

Fuchsian Groups

10.1 Introduction

In this chapter we will specialize to the case $n = 2$. A discrete group preserving the unit ball (disc) B in R^2 is a Fuchsian group. We are singling out this special class of groups for special attention because it will give us the opportunity to do two things.

1. We give a different presentation of the line element space, flows on this space, and measures invariant under the flow. In contrast to our previous derivation, which was entirely geometric, we adopt an algebraic approach. This has the advantage of yielding algebraic formulae for the flows which can be handled with more facility than the somewhat descriptive geometric notions. Our approach is modeled on the work of Fomin and Gel'fand [Fomin and Gel'fand, 1952].

2. We introduce the horocycle flow. This flow is closely related to the geodesic flow and we will exploit the relationship to derive ergodic properties of both flows. Additionally we will find that the horocycle flow enjoys properties (minimality, unique ergodicity) which the geodesic flow does not. The horocycle flow can be defined in n-dimensions, indeed it was introduced in this generality by Hopf [Hopf, 1939], and has been studied by many authors. However, it is simplest, and in some senses most natural, to introduce it in the two-dimensional setting.

We will present an account of ergodic phenomena for Fuchsian groups which is essentially independent of the treatment in chapter 8. This has the advantage of making the chapter self contained and, more importantly, of demonstrating proofs of the ergodic theorems which exploit the connection between the geodesic and horocyclic flows. As always in the study of discrete groups, we have an upper half-plane model, and a disc model. Each has its own advantages and they are sufficiently different to warrant separate sections.

10.2 The Upper Half-Plane

We define $M(H)$ to be the group of Moebius transformations preserving the upper half, H, of the complex plane. Thus

$$M(H) = \left\{ \frac{az + b}{cz + d} : a,b,c,d \in R, \ ad - bc = 1 \right\}.$$

We remark that $M(H)$ is isomorphic to $SL(2,R) / \pm I$. We have a metric on $M(H)$ derived from the norm

$$\left\| g \right\| = \left\| \frac{az + b}{cz + d} \right\| = \left(a^2 + b^2 + c^2 + d^2 \right)^{1/2}$$

and $M(H)$ is a topological group with respect to this metric.

We wish to put a measure μ_H on $M(H)$ and this is done by means of the differential

$$d\mu_H(g) = \frac{d\beta \, d\gamma \, d\delta}{|\delta|} \quad where \quad g = \begin{pmatrix} \alpha & \beta \\ \gamma & \delta \end{pmatrix}$$

— from now on we will frequently identify the Moebius transform with its associated matrix. Let $g = \begin{pmatrix} \alpha & \beta \\ \gamma & \delta \end{pmatrix}$, $h = \begin{pmatrix} a & b \\ c & d \end{pmatrix}$ and write $g' = h \circ g = \begin{pmatrix} \alpha' & \beta' \\ \gamma' & \delta' \end{pmatrix}$ then we have

$$\beta' = a\beta + b\delta, \ \gamma' = \frac{c(1 + \beta\gamma)}{\delta} + d\gamma, \ \delta' = c\beta + d\delta$$

and the Jacobian of the transformation $(\beta,\gamma,\delta) \rightarrow (\beta',\gamma',\delta')$ is easily calculated to be δ'/δ. It follows that

$$\frac{d\beta' \, d\gamma' \, d\delta'}{|\delta'|} = \frac{d\beta \, d\gamma \, d\delta}{|\delta|},$$

and we have shown that μ_H is invariant under left composition. An entirely similar proof shows its invariance under right composition. A Borel measure on a topological group which is both left and right invariant is a Haar measure, and we have proved.

Theorem 10.2.1   The measure μ_H on $M(H)$ defined by

$$d\mu_H(g) \;=\; \frac{d\beta\,d\gamma\,d\delta}{|\delta|}\quad where\quad g = \begin{pmatrix} \alpha & \beta \\ \gamma & \delta \end{pmatrix}$$

is the Haar measure on $M(H)$.

We consider three subgroups of $M(H)$ as follows :

$$K \;=\; \left\{ \begin{pmatrix} \cos\theta/2 & \sin\theta/2 \\ -\sin\theta/2 & \cos\theta/2 \end{pmatrix} : \theta \in R \right\}$$

which is the stabilizer subgroup of i. Note that the transform in K above is (at i) a counter clockwise rotation through θ.

$$A \;=\; \left\{ \begin{pmatrix} y^{1/2} & 0 \\ 0 & y^{-1/2} \end{pmatrix} : y > 0 \right\}$$

is the group of homothetic transformations, and

$$N \;=\; \left\{ \begin{pmatrix} 1 & x \\ 0 & 1 \end{pmatrix} : x \in R \right\}$$

which is the group of translations.

If $g \in M(H)$ with $g(i) = x + iy$ and $arg(g'(i)) = \theta$ then an easy calculation shows that g may be written uniquely as the product

$$g \;\equiv\; \begin{pmatrix} 1 & x \\ 0 & 1 \end{pmatrix} \begin{pmatrix} y^{1/2} & 0 \\ 0 & y^{-1/2} \end{pmatrix} \begin{pmatrix} \cos\theta/2 & \sin\theta/2 \\ -\sin\theta/2 & \cos\theta/2 \end{pmatrix} \tag{10.2.1}$$

for $\theta \in [0,2\pi)$. This corresponds of course to the Iwasawa decomposition of $SL(2,R)$ given by

$$\begin{pmatrix} a & b \\ c & d \end{pmatrix} = \begin{pmatrix} 1 & x \\ 0 & 1 \end{pmatrix} \begin{pmatrix} y^{1/2} & 0 \\ 0 & y^{-1/2} \end{pmatrix} \begin{pmatrix} \cos\theta & \sin\theta \\ -\sin\theta & \cos\theta \end{pmatrix}$$

which is a unique decomposition for θ in the range $[0,\pi)$. It will be very useful to have the form of the invariant measure $d\mu_H$ in terms of (x,y,θ) and from a routine calculation,

$$\beta = y^{1/2}\sin\theta/2 + x\,y^{-1/2}\cos\theta/2,\quad \gamma = -y^{-1/2}\sin\theta/2,\quad \delta = y^{-1/2}\cos\theta/2.$$

From this the Jacobian of the transformation $(x,y,\theta) \to (\beta,\gamma,\delta)$ is calculated to be $(-\cos(\theta/2))/4y^{5/2}$, and so

$$d\mu_H \;=\; \frac{d\beta\,d\gamma\,d\delta}{|\delta|} \;=\; \frac{dx\,dy\,d\theta}{4y^2}. \tag{10.2.2}$$

We remark in passing that, since the upper half-plane may be identified, via

(10.2.1), with $M(H)/K$, then (10.2.2) provides another proof of the fact that $\dfrac{dx\,dy}{4y^2}$ is an invariant area measure on the upper half-plane.

We next introduce line elements. A **line element** is defined to be a triple (x,y,θ) with x,y,θ real, $y > 0$ and $0 \leq \theta < 2\pi$. The space Ω_H of line elements inherits the Euclidean topology as a subset of R^3 and is in fact homeomorphic to $M(H)$ — the Moebius group on the upper half-plane — via the map $\phi_H : \Omega_H \to M(H)$ defined by

$$\phi_H(x,y,\theta) \;=\; \begin{pmatrix} 1 & x \\ 0 & 1 \end{pmatrix} \begin{pmatrix} y^{1/2} & 0 \\ 0 & y^{-1/2} \end{pmatrix} \begin{pmatrix} \cos\theta/2 & \sin\theta/2 \\ -\sin\theta/2 & \cos\theta/2 \end{pmatrix}.$$

A group operation $*$ may be defined on Ω_H by

$$(x,y,\theta)*(u,v,\alpha) \;=\; \phi_H^{-1}\left[\phi_H(x,y,\theta)\circ\phi_H(u,v,\alpha)\right]$$

which makes ϕ_H a group homomorphism and gives Ω_H the structure of a topological group. On Ω_H we define the measure M by

$$dM(x,y,\theta) \;=\; \frac{dx\,dy\,d\theta}{y^2}.$$

This is, of course, precisely the measure M introduced in section 8.1. The action of $M(H)$ on Ω_H can be defined in two distinct ways and we wish to consider both of these. For $h \in M(H)$ we define

$$h(x,y,\theta) \;=\; \phi_H^{-1}\left[h\circ\phi_H(x,y,\theta)\right] \tag{10.2.3}$$

and

$$h_R(x,y,\theta) \;=\; \phi_H^{-1}\left[\phi_H(x,y,\theta)\circ h\right]. \tag{10.2.4}$$

We next show that the measure M defined above is invariant under both actions of $M(H)$ on Ω_H. To this end suppose $(x',y',\theta') = h(x,y,\theta)$ then $\phi_H(x',y',\theta') = h\circ\phi_H(x,y,\theta)$ and so if we write

$$\phi_H(x',y',\theta') \;=\; \begin{pmatrix} \alpha' & \beta' \\ \gamma' & \delta' \end{pmatrix} \quad \text{and} \quad \phi_H(x,y,\theta) \;=\; \begin{pmatrix} \alpha & \beta \\ \gamma & \delta \end{pmatrix}$$

then

$$\begin{pmatrix} \alpha' & \beta' \\ \gamma' & \delta' \end{pmatrix} \;=\; h\circ\begin{pmatrix} \alpha & \beta \\ \gamma & \delta \end{pmatrix}$$

and $\dfrac{d\beta'\,d\gamma'\,d\delta'}{|\delta'|} = \dfrac{d\beta\,d\gamma\,d\delta}{|\delta|}$ from theorem 10.2.1. However, from (10.2.2),

$$\frac{dx'\,dy'\,d\theta'}{4(y')^2} = \frac{d\beta'\,d\gamma'\,d\delta'}{|\delta'|} \quad \text{and} \quad \frac{dx\,dy\,d\theta}{4y^2} = \frac{d\beta\,d\gamma\,d\delta}{|\delta|}$$

and we have shown the invariance of M under the action (10.2.3). The proof for the right action is entirely similar.

Theorem 10.2.2   The measure M on Ω_H defined by

$$dM = \frac{dx\,dy\,d\theta}{y^2} \quad \text{at} \quad (x,y,\theta)$$

is invariant under both actions (10.2.3) and (10.2.4) of $M(H)$ on Ω_H.

We next want to realize line elements and the group action on them in geometric terms. The line element (x,y,θ) should be viewed as the point $x + iy$ of the upper half-plane together with a direction making an angle θ with the upward vertical at $x + iy$.

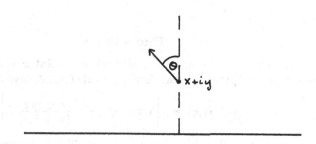

Figure 10.2.1

Starting with the line element (x,y,θ) define the Moeblus transform

$$g = \begin{pmatrix} 1 & x \\ 0 & 1 \end{pmatrix} \begin{pmatrix} y^{1/2} & 0 \\ 0 & y^{-1/2} \end{pmatrix} \begin{pmatrix} \cos\theta/2 & \sin\theta/2 \\ -\sin\theta/2 & \cos\theta/2 \end{pmatrix}$$

and we verify that the action of g on the point i is to move it to the point $x + iy$, and that the vertical direction at i is mapped to the direction making an angle θ with the upward vertical at $x + iy$. Thus in terms of the action of a Moebius transform on points and directions we would expect that

$$g(0,1,0) = (x,y,\theta).$$

This is in fact the case, and follows from the definition (10.2.3) because

$$g(0,1,0) = \phi_H^{-1}(g \circ \phi_H(0,1,0)) = \phi_H^{-1}(g)$$

since $\phi_H(0,1,0)$ is the identity, and we note that $\phi_H^{-1}(g) = (x,y,\theta)$. We have the following situation. Given a line element (x,y,θ) there is associated a unique

Moebius transformation $g = \phi_H(x,y,\theta)$ such that $g(0,1,0) = (x,y,\theta)$. In other words, $g(i) = x + iy$ and $arg(g'(i)) = \theta$. Now what is the action of a Moebius transform h on the line element (x,y,θ) ? We verify immediately from (10.2.3) that h moves the point $x + iy$ to $h(x + iy)$ and the angle θ is increased by an amount $arg(h'(x + iy))$. In other words h acts according to (10.2.3) exactly as one would expect a Moebius transform (or indeed any analytic function) to act on a point and a direction.

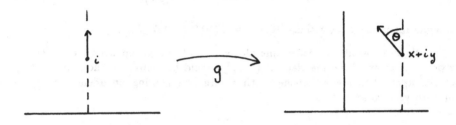

Figure 10.2.2

Put another way, if we regard a line element as a point $x + iy$ together with a unit vector $\xi \ (= e^{i\theta})$ then the transformation rule (10.2.3) says that

$$h(x + iy, \xi) = \left[h(x + iy), \frac{h'(x + iy)}{|h'(x + iy)|} \xi \right].$$

This is exactly the action defined in section 8.1. The transformation rule (10.2.4) is of an entirely different character, it is the one we use in the next section to define the geodesic and horocyclic flows, and we defer discussion of it until that time. For the remainder of this section we are concerned with placing different coordinate systems on the space Ω_H and computing the invariant volume element dM in these new coordinates. There are three different coordinate systems of interest.

For the first we proceed as follows. Given (x,y,θ) we determine the geodesic passing through $x + iy$ in the direction θ and let ξ be the end point. We can clearly write the line element as the triple (x,y,ξ) and can in fact calculate that

$$\xi = x - y \cot(\theta/2).$$

The Jacobian of the transformation $(x,y,\theta) \to (x,y,\xi)$ is seen to be $y/(2\sin^2(\theta/2))$, and so

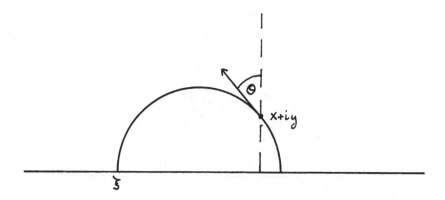

Figure 10.2.3

$$dM = \frac{dx\ dy\ d\theta}{y^2} = 2\sin^2(\theta/2)\frac{dx\ dy\ d\xi}{y^3}$$

$$= \frac{2\ dx\ dy\ d\xi}{y\,|(x+iy)-\xi|^2}$$

which is the invariant measure element in (x,y,ξ) coordinates. The group action in terms of these coordinates is found to be

$$h(x+iy,\ \xi) = (h(x+iy),\ h(\xi)).$$

It is further to be noted that if

$$\phi_H(x,y,\theta) = g$$

then $g(i) = x + iy$ and $g(\infty) = \xi$.

Our second coordinate set is obtained as follows. Given (x,y,θ), find the geodesic passing through $x + iy$ in the direction θ and let η, ξ be the beginning and end points respectively. Now let s be the (directed) distance of $x + iy$ from the highest point of the geodesic. This is precisely the coordinate set introduced in section 8.1. We already know from section 8.1 the invariant measure in (η,ξ,s) coordinates, but it is of interest to recompute it via the Haar measure. Let g be the Moebius transform $\phi_H(x,y,\theta)$ and note that $g(0) = \eta$, $g(i) = x + iy$, $g(\infty) = \xi$. Then writing

$$g = \begin{pmatrix} \alpha & \beta \\ \gamma & \delta \end{pmatrix}$$

we have $\eta = \beta/\delta$ and $\xi = \alpha/\gamma$. We need to calculate the distance s in terms of

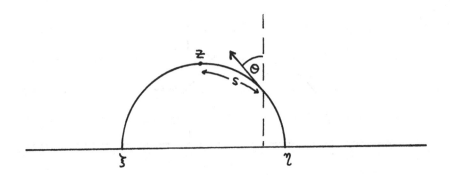

Figure 10.2.4

$\alpha,\beta,\gamma,\delta$. In order to do this, suppose $iu = g^{-1}(z)$ then

$$\rho(x + iy, z) = \rho(g^{-1}(x + iy), g^{-1}(z))$$

$$= \rho(i, iu) = \log(|u|).$$

Note that iu is the point on the imaginary axis whose g-image has the largest imaginary part. But

$$\text{Im } g(iv) = \frac{v}{\delta^2 + \gamma^2 v^2}$$

which is maximized when $v = |\delta|/|\gamma|$. Thus $u = |\delta|/|\gamma|$ and $s = \log(|\delta|/|\gamma|)$, and we find

$$\frac{\partial(\eta,\xi,s)}{\partial(\beta,\gamma,\delta)} = \begin{vmatrix} \dfrac{1}{\delta} & 0 & -\dfrac{\beta}{\delta^2} \\ \dfrac{1}{\delta} & \dfrac{-1}{\delta\gamma^2} & \dfrac{-(1+\beta\gamma)}{\gamma\delta^2} \\ 0 & \dfrac{-1}{\gamma} & \dfrac{1}{\gamma} \end{vmatrix} = \frac{-2}{\delta^3\gamma^2},$$

and so $d\eta\, d\xi\, ds = \dfrac{2d\beta\, d\gamma\, d\delta}{|\delta|\delta^2\gamma^2}$. But we have $|g(0) - g(\infty)|^2 = \dfrac{1}{\delta^2\gamma^2}$ and so

$$\frac{d\eta\, d\xi\, ds}{|\eta - \xi|^2} = \frac{2d\beta\, d\gamma\, d\delta}{|\delta|}$$

is the invariant measure element in (η,ξ,s) coordinates.

Our final representation of Ω is obtained as follows. Given (x,y,θ), determine the geodesic as before with ξ as its end point. Now construct a horocycle at ξ passing through $x + iy$. Let r be the Euclidean radius of the horocycle and s the directed (hyperbolic) distance measured along the horocycle from $x + iy$ to the highest point of the horocycle.

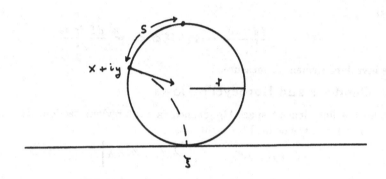

Figure 10.2.5

The triple (ξ,r,s) determines the line element (x,y,θ). Let g be the Moebius transform $\phi_H(x,y,\theta)$ and note that $g(i) = x + iy$ and $g(\infty) = \xi$. Thus $\xi = \alpha/\gamma = (1 + \beta\gamma)/(\gamma\delta)$. We need to represent r and s in terms of β,γ,δ. Note that the horocycle H is the g-image of the horocycle $\{z : \mathrm{Im}\, z = 1\}$. Thus the highest point of H is the point $g(u + i)$ which has the largest imaginary part. Now

$$\mathrm{Im}\, g(u + i) = \frac{1}{(\gamma u + \delta)^2 + \gamma^2}$$

which is maximized when $u = -\delta/\gamma$ and has a maximum value of $1/\gamma^2$. Thus $r = 1/(2\gamma^2)$ and to compute s we note that g preserves hyperbolic distance, and we need only measure the distance from i to $i - \delta/\gamma$ along the horocycle $\{z : \mathrm{Im}\, z = 1\}$. This distance is just δ/γ and we have

$$\xi = \frac{1 + \beta\gamma}{\gamma\delta}, \quad r = \frac{1}{2\gamma^2}, \quad s = \frac{\delta}{\gamma}.$$

We calculate

$$\frac{\partial(\xi,r,s)}{\partial(\beta,\gamma,\delta)} \;=\; \begin{vmatrix} \dfrac{1}{\delta} & \dfrac{-1}{\delta\gamma^2} & \dfrac{-(1+\beta\gamma)}{\gamma\delta^2} \\[2mm] 0 & \dfrac{-1}{4\gamma^3} & 0 \\[2mm] 0 & \dfrac{-\delta}{\gamma^2} & \dfrac{1}{\gamma} \end{vmatrix} \;=\; \frac{-1}{4\delta\gamma^4}$$

and so

$$dM \;=\; \frac{d\beta\,d\gamma\,d\delta}{|\delta|} \;=\; 4\gamma^4\,d\xi\,dr\,ds \;=\; \frac{d\xi\,dr\,ds}{r^2}$$

is the invariant element of measure.

10.3 Geodesic and Horocyclic Flows

Consider the line element space Ω_H defined in the previous section. Recall that for $h \in M(H)$ we have defined the right action

$$h_R(x,y,\theta) \;=\; \phi_H^{-1}\!\left[\phi_H(x,y,\theta)\circ h\right].$$

We will only use this action for three special types of transformation. For t real the geodesic flow g_t and the horocyclic flow h_t are defined on Ω_H by

$$g_t(x,y,\theta) \;=\; \begin{pmatrix} e^{t/2} & 0 \\ 0 & e^{-t/2} \end{pmatrix}_R (x,y,\theta)$$

and

$$h_t(x,y,\theta) \;=\; \begin{pmatrix} 1 & t \\ 0 & 1 \end{pmatrix}_R (x,y,\theta).$$

The rotation R_α is defined on Ω_H by

$$R_\alpha(x,y,\theta) \;=\; \begin{pmatrix} \cos\alpha/2 & \sin\alpha/2 \\ -\sin\alpha/2 & \cos\alpha/2 \end{pmatrix}_R (x,y,\theta),$$

and it is routine to check that $R_\alpha(x,y,\theta) = (x,y,\theta+\alpha)$. Note that

$$g_t(x,y,\theta) \;=\; \phi_H^{-1}\!\left[\phi_H(x,y,\theta)\begin{pmatrix} e^{t/2} & 0 \\ 0 & e^{-t/2} \end{pmatrix}\right]$$

$$h_t(x,y,\theta) \;=\; \phi_H^{-1}\!\left[\phi_H(x,y,\theta)\begin{pmatrix} 1 & t \\ 0 & 1 \end{pmatrix}\right]$$

and so $g_t(x,y,\theta) = (x,y,\theta)$ if and only if $t = 0$. Further, $h_t(x,y,\theta) = (x,y,\theta)$ if and only if $t = 0$. For any t,s real, $g_s \circ g_t \equiv g_{s+t}$ and $h_s \circ h_t \equiv h_{s+t}$. Thus in each case we have a one parameter group of transformations acting on the space Ω and, from theorem 10.2.2, these transforms preserve the measure M on Ω_H.

We wish to consider the geometric action of these flows. For g_t let $u = \phi_H(x,y,\theta)$ with $u(0) = \eta$, $u(\infty) = \xi$. Clearly, if $v = \phi(g_t(x,y,\theta))$ then $v(0) = \eta$ and $v(\infty) = \xi$. In other words (x,y,θ) and $g_t(x,y,\theta)$ determine the same geodesic.

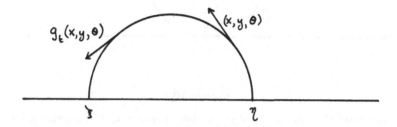

Figure 10.3.1

The point (x,y,θ) is obtained as $u(0,1,0)$ whereas $g_t(x,y,\theta)$ is $u(0,e^t,0)$ and so the hyperbolic distance between the base points of the elements (x,y,θ) and $g_t(x,y,\theta)$ is the same as that between i and $e^t i$. This distance is t. Thus g_t represents a flow, through a directed distance t, along the geodesic determined by a line element. In terms of the coordinates (η,ξ,s) introduced in the previous section we have

$$g_t(\eta,\xi,s) = (\eta,\xi,s+t).$$

What about the flow h_t? If we write $u = \phi(x,y,\theta)$ with $u(\infty) = \xi$ then, writing $v = \phi_H(h_t(x,y,\theta))$, we have $v(\infty) = \xi$. The two line elements thus determine geodesics ending at the same point ξ. Since the transform $z \rightarrow z + t$ preserves the line $\{z : \mathrm{Im}\, z = 1\}$ we note further that (x,y,θ) and $h_t(x,y,\theta)$ are both line elements which are based on, and orthogonal to, the horocycle at ξ, $X = u(\{z : \mathrm{Im}\, z = 1\})$. The distance, measured on the horocycle X, between the base points of the two elements is equal to the distance, along $\{z : \mathrm{Im}\, z = 1\}$, between i and $i + t$. This distance is t. Thus h_t represents a flow, through a directed distance t, along the horocycle determined by a line element. In terms of the coordinates (ξ,r,s) introduced in the previous section we have

$$h_t(\xi,r,s) = (\xi,r,s + t).$$

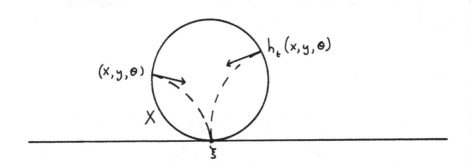

Figure 10.3.2

We conclude this section by deriving two important relationships between the flows. The first relation is suggested by figure 10.3.3.

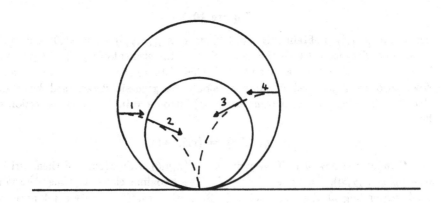

Figure 10.3.3

Four line elements are shown. A horocyclic flow from (1) to (4) could be accomplished instead in three stages — a geodesic flow from (1) to (2), a horocyclic flow from (2) to (3), and a geodesic flow from (3) to (4). Thus, given s and t, we would expect to find reals u, v such that

$$g_s \circ h_t = h_u \circ g_v.$$

It is easy to show that $v = s$ and not too difficult to compute u. However, we prefer the algebraic approach. Note that

$$\begin{pmatrix} 1 & t \\ 0 & 1 \end{pmatrix} \begin{pmatrix} e^{s/2} & 0 \\ 0 & e^{-s/2} \end{pmatrix} = \begin{pmatrix} e^{v/2} & 0 \\ 0 & e^{-v/2} \end{pmatrix} \begin{pmatrix} 1 & u \\ 0 & 1 \end{pmatrix}$$

if and only if $s = v$ and $u = te^{-s}$. This proves the following result.

Theorem 10.3.1   The flows g, h on Ω_H are related by :

$$g_s \circ h_t \equiv h_{te^{-s}} \circ g_s.$$

There is another relation between the flows and it is suggested by figure 10.3.4.

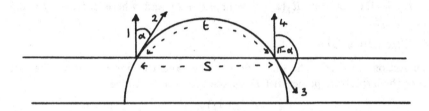

Figure 10.3.4

Starting with the line element (1) , rotate through $2\pi - \alpha$ to obtain element number (2). Now apply the geodesic flow g_t to arrive at (3). Another way to arrive at (3) starting from (1) is first to apply the horocyclic flow h_s to get (4) and then rotate through $\pi + \alpha$. Note that a rotation through $2\pi - \alpha$ is represented by the matrix

$$\begin{pmatrix} -\cos\alpha/2 & \sin\alpha/2 \\ -\sin\alpha/2 & -\cos\alpha/2 \end{pmatrix}.$$

Whereas a rotation through $\pi + \alpha$ is represented by

$$\begin{pmatrix} -\sin\alpha/2 & \cos\alpha/2 \\ -\cos\alpha/2 & -\sin\alpha/2 \end{pmatrix}.$$

Given s then, we wish to solve the system

$$\begin{pmatrix} -\cos\alpha/2 & \sin\alpha/2 \\ -\sin\alpha/2 & -\cos\alpha/2 \end{pmatrix} \begin{pmatrix} e^{t/2} & 0 \\ 0 & e^{-t/2} \end{pmatrix} = \begin{pmatrix} 1 & s \\ 0 & 1 \end{pmatrix} \begin{pmatrix} -\sin\alpha/2 & \cos\alpha/2 \\ -\cos\alpha/2 & -\sin\alpha/2 \end{pmatrix}$$

which reduces to the system

$$e^{t/2}\cos\alpha/2 = \sin\alpha/2 + s\cos\alpha/2$$

$$e^{-t/2}\sin\alpha/2 = \cos\alpha/2 - s\sin\alpha/2$$

$$e^{t/2}\sin\alpha/2 = \cos\alpha/2.$$

Thus $e^{t/2} = \cot\alpha/2$ and from the first equation we find $\cot\alpha = s/2$. Now solve for t to find $s = 2\sinh(t/2)$ and we have proved the next result.

Theorem 10.3.2   The flows g, h on Ω_H are related by

$$h_s \equiv R_{\pi-\alpha} \circ g_t \circ R_{2\pi-\alpha}$$

where R_λ is the rotation $R_\lambda(x,y,\theta) = (x,y,\theta+\lambda)$ and where $\cot\alpha = s/2$ and $s = 2\sinh(t/2)$.

10.4 The Unit Disc

In this section we derive the line element space and the flows on it in the disc model of the hyperbolic plane. Let D denote the open unit disc

$$D = \{z : |z| < 1\}$$

and let $M(D)$ denote the Moebius group preserving D. If $T \in M(D)$ then we have

$$T(z) = \left\{ \frac{az+\bar{c}}{cz+\bar{a}} : |a|^2 - |c|^2 = 1 \right\}.$$

A line element in D is a triple of reals (u,v,α) with $u^2 + v^2 < 1$ and $0 \le \alpha < 2\pi$. Such an element is to be viewed as a point $u + iv$ of D together with a direction making the angle α with the positive real axis — see figure 10.4.1. We write Ω_D for the space of all line elements. A Moebius transform acts on Ω_D in the obvious way — namely

$$T(u,v,\alpha) = (\operatorname{Re} T(u+iv), \operatorname{Im} T(u+iv), \alpha \dotplus \arg T'(u+iv)).$$

Where $\dotplus$ denotes addition modulo 2π.

Exactly as in the upper half-plane, a line element in D yields a directed geodesic and a horocycle as shown in figure 10.4.2. We can use these ideas to derive a coordinate system on Ω_D which is much more natural for our purposes

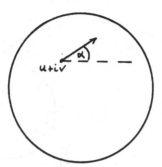

Figure 10.4.1

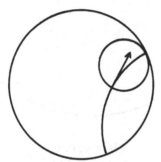

Figure 10.4.2

than the (u,v,α) system. With this new system the flows are easily and naturally
defined and a measure invariant under both Moebius action and the flows is fairly
easily derived. This work constitutes the bulk of this section, however, we will
also consider some other coordinates on Ω_D which have appeared in the literature
and develop the form of the invariant measure in these coordinates.

Consider the line element (u,v,α) and let $\xi\ (=e^{i\theta})$ be the end point of the
directed geodesic determined by this element. Now construct the horocycle A
determined by the element and write w for the point where A meets the diameter
to ξ. Let t be the directed hyperbolic distance from 0 to w (measured towards ξ)
and write s for the directed hyperbolic distance from w to $u+iv$ measured
clockwise around A. Figure 10.4.3 shows these quantities. The triple (θ,t,s)
determines the line element (u,v,α) uniquely and we use these new coordinates to
derive our flows and the invariant measure. We start by finding a measure on Ω_D

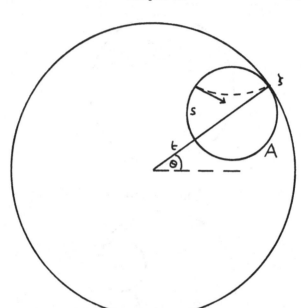

Figure 10.4.3

which is invariant under Moebius action.

Theorem 10.4.1   A measure on Ω_D which is invariant under the action of $M(D)$ on Ω_D is given by the element

$$dm_D \;=\; e^t\, d\theta\, dt\, ds\;.$$

Proof. Suppose $T \in M(D)$ and $(\theta, t, s) \in \Omega_D$. We write $(\theta', t', s') = T(\theta, t, s)$ and we must therefore compute the Jacobian of the map. Now $e^{i\theta'} = T(e^{i\theta})$ and so $\partial\theta'/\partial t = \partial\theta'/\partial s = 0$. It is geometrically evident that t' is independent of s and so $\partial t'/\partial s = 0$. Thus

$$\frac{\partial(\theta', t', s')}{\partial(\theta, t, s)} \;=\; \left|\; \frac{d\theta'}{d\theta}\cdot\frac{\partial t'}{\partial t}\cdot\frac{\partial s'}{\partial s}\;\right|\;.$$

Evidently $d\theta'/d\theta = (T'(e^{i\theta})\, e^{i\theta})/T(e^{i\theta})$ which must be real (since θ and θ' are real) and so

$$\left|\; \frac{d\theta'}{d\theta}\;\right| \;=\; |T'(e^{i\theta})|\;. \qquad\qquad (10.4.1)$$

Writing r for the Euclidean radius of A and r' for that of $T(A)$ we recall from lemma 2.5.2 that

$$r' = \frac{r\,|T'(e^{i\theta})|}{1 - r + r\,|T'(e^{i\theta})|}$$

in other words

$$\frac{r'}{1 - r'} = T'(e^{i\theta})\,\frac{r}{1 - r}\ .$$

However, a routine calculation shows that $t = \log\left((1 - r)/r\right)$ and so

$$e^{-t'} = |T'(e^{i\theta})|e^{-t}\ . \tag{10.4.2}$$

Thus $t' = t - \log|T'(e^{i\theta})|$ and so

$$\frac{\partial t'}{\partial t} = 1\ . \tag{10.4.3}$$

Considering figure 10.4.4 it is clear that $s' = s + d$ where d measures the distance along the horocycle $T(A)$ from the point α to $T(w)$.

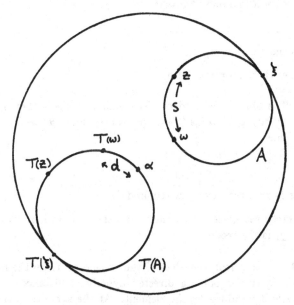

Figure 10.4.4

This distance d depends only on ξ, t and T and so $\partial d/\partial s = 0$. Thus it follows that

$$\frac{\partial s\,'}{\partial s} = 1 \quad . \tag{10.4.4}$$

From equations (10.4.1), (10.4.2), (10.4.3), and (10.4.4) we see that

$$\frac{\partial(\theta',t\,',s\,')}{\partial(\theta,t,s)} = |T\,'\,(e^{i\theta})| = e^{t\,-\,t\,'}$$

which proves the theorem. □

The advantage of this particular coordinate system is the ease with which the flows may be defined. For v real define maps g_v and h_v from Ω_D onto itself by

$$h_v\,(\theta,t,s) = (\theta,t,s+v) \quad \text{and} \quad g_v\,(\theta,t,s) = (\theta,t+v,se^{-v}) \quad .$$

Let us first show that these maps are flows.

Theorem 10.4.2   For any real v the maps h_v, g_v are continuous one-to-one maps of Ω_D onto itself. Further

1. h_v is the identity if and only if $v = 0$

2. $h_v \circ h_u = h_{v+u}$

3. h_v preserves the measure m_D

The map g_v satisfies these properties also.

Proof. This result is almost self evident. We shall just check that g_v preserves the measure m_D. Writing $(\theta',t\,',s\,') = g_v(\theta,t,s)$ we have $\theta' = \theta$, $t\,' = t+v$, and $s\,' = se^{-v}$. Thus

$$\frac{\partial(\theta',t\,',s\,')}{\partial(\theta,t,s)} = e^{t-t\,'}$$

and so $e^t\,d\theta\,dt\,ds$ is preserved by g_v as required. □

We immediately recognize h_v as the horocycle flow and in view of figure 10.4.5 we see that g_v is the geodesic flow.

Theorem 10.4.3   For v real, the effect of the transform h_v on the line element (θ,t,s) is to move its carrier point a directed hyperbolic distance v (clockwise) around the horocycle determined by the element. At the same time the direction stays internally normal to the horocycle.

For v real the effect of the transform g_v on the line element (θ,t,s) is to move its carrier point a directed hyperbolic distance v along the directed geodesic determined by the line element. At the same time the direction stays tangent to

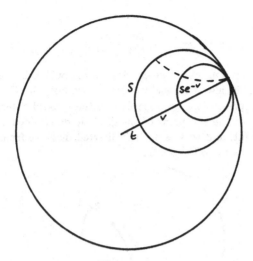

Figure 10.4.5

the geodesic.

The next result is evident from figure 10.4.5, and also can be verified immediately from the definitions.

Theorem 10.4.4   For u,v real

$$g_u \circ h_v \; = \; h_{ve^{-\imath}} \circ g_u \; .$$

We next consider two other coordinate systems on Ω_D. The first system is the one with which we started the section — namely (u,v,α). Now if $T \in M(D)$ we have already seen that

$$T(u,v,\alpha) \; = \; (Re \; T(u+iv), Im \; T(u+iv), \alpha + arg T'(u+iv)) \; = \; (u',v',\alpha')$$

say. It is clear that $arg \; T'(u+iv)$ is independent of α and so $\partial\alpha'/\partial\alpha = 1$. Since we know that

$$\frac{\partial(u',v')}{\partial(u,v)} \; = \; \frac{[1-(u')^2-(v')^2]^2}{[1-u^2-v^2]^2}$$

it follows that the element

$$dm \; = \; \frac{du \; dv \; d\alpha}{[1-(u^2+v^2)]^2}$$

is invariant under Moebius action on Ω_D. The Radon-Nikodym derivative dm_D/dm is thus a function on Ω_D invariant under $M(D)$ — and so reduces to a constant. Thus there is a constant k such that

$$dm_D = \frac{k \, du \, dv \, d\alpha}{[1 - (u^2 + v^2)]^2} \, .$$

The other coordinate system which we consider is one which figures prominently in the literature and is also the one we used previously in Chapter 8. It is particularly easy to represent the geodesic flow in these coordinates. Given a line element (u, v, α) let ξ be the end point of the geodesic determined by the element, let η be the initial point, and write s for the directed distance from $u + iv$ to the point of the geodesic closest to the origin.

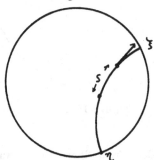

Figure 10.4.6

The line element may be written as the triple (η, ξ, s). Clearly the geodesic flow acts by the rule $g_v(\eta, \xi, s) = (\eta, \xi, s + v)$. We wish to determine the measure m_D in terms of the coordinates (η, ξ, s) and to this end we set $\eta = e^{i\phi}$, $\xi = e^{i\theta}$, choose $T \in M(D)$ and note that if

$$(e^{i\phi'}, e^{i\theta'}, s') = T(e^{i\phi}, e^{i\theta}, s)$$

then $e^{i\phi'} = T(e^{i\phi})$ and $e^{i\theta'} = T(e^{i\theta})$. Thus

$$\frac{\partial \phi'}{\partial \theta} = \frac{\partial \phi'}{\partial s} = \frac{\partial \theta'}{\partial \phi} = \frac{\partial \theta'}{\partial s} = 0$$

and so

$$\frac{\partial(\phi', \theta', s)}{\partial(\phi, \theta, s)} = \frac{\partial \phi'}{\partial \phi} \cdot \frac{\partial \theta'}{\partial \theta} \cdot \frac{\partial s'}{\partial s}. \tag{10.4.5}$$

In figure 10.4.7 we observe that $s' = s + d$ where, if z is the mid point of the geodesic joining η and ξ and w is the mid point of the geodesic joining $T(\eta)$ and $T(\xi)$,

$$d = \rho(w, T(z)).$$

Thus d depends only on ξ, η and T so $\partial d/\partial s = 0$ and $\partial s'/\partial s = 1$. From (10.4.5) we have

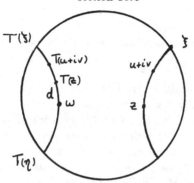

Figure 10.4.7

$$\frac{\partial(\phi',\theta',s')}{\partial(\phi,\theta,s)} = \frac{\partial\phi'}{\partial\phi} \cdot \frac{\partial\theta'}{\partial\theta}.$$

Now $e^{i\phi'} = T(e^{i\phi})$ and so

$$\frac{d\phi'}{d\phi} = T'(e^{i\phi})\frac{e^{i\phi}}{T(e^{i\phi})}$$

and this must be real (since ϕ, ϕ' are real) thus

$$\frac{d\phi'}{d\phi} = \pm|T'(e^{i\phi})|$$

with a similar result for $d\theta'/d\theta$. Thus

$$\frac{\partial(\phi',\theta',s')}{\partial(\phi,\theta,s)} = |T'(e^{i\phi})||T'(e^{i\theta})|$$

$$= \frac{|T(e^{i\phi}) - T(e^{i\theta})|^2}{|e^{i\phi} - e^{i\theta}|^2}$$

and we have shown that the measure M defined on Ω_D by

$$dM = \frac{d(arg\,\xi)\,d(arg\,\eta)\,ds}{|\xi - \eta|^2}$$

is invariant under the action of $M(D)$ on Ω_D. The Radon-Nikodym derivative dM/dm_D is thus a function on Ω_D which is invariant under $M(D)$. But $M(D)$ acts transitively on Ω_D and so this function is constant. Thus, for some constant k,

$$dm_D = \frac{kd(arg\,\xi)\,d(arg\,\eta)\,ds}{|\xi - \eta|^2}.$$

The measures we have introduced so far on the line element space — both in the disc and in the upper half-plane — are all variations on the same theme, and are all the same up to a constant multiple. We shall consistently use the letter M for these measures no matter what coordinate system is in use.

10.5 Ergodicity and Mixing

In this section we will consider ergodicity and mixing for the geodesic and horocyclic flows on the quotient space with respect to the measure M. In view of the fact that g_t and h_s both commute with Moebius transforms (this was proved in section 8.1), they both have an action on the quotient space Ω_H / Γ or Ω_D / Γ. In this section and the next we will use the symbol Ω for the quotient line element space. The measure M leads to a measure on Ω as outlined in Chapter 8. This measure will also be denoted M. Similarly, the metric d on Ω is defined just as in section 8.1.

In this section we consider Fuchsian groups of finite area and aim to show that both the horocyclic and geodesic flows are mixing. All results given are in relation to the measure M. To be precise, the words ergodic and mixing, when used in this section, mean "ergodic (M)" and "mixing (M)".

It is important to note that the methods of this section cannot be applied to prove ergodicity and mixing properties relative to the measure m_z derived from a conformal density as in section 8.1. The reason for this is simply that the measure m_z is not invariant under the horocyclic flow. In fact, we shall see in the next section that, for many groups, M is the **only** measure invariant under the horocyclic flow. It turns out that the geodesic flow does enjoy a form of mixing (weak mixing) but the proof of this fact is entirely different from the methods of this section and will not be covered in this book. The interested reader is referred to the work of Rudolph [Rudolph, 1982] for a full account.

Much of the work in this section dates to the 1930's. For example, the ergodicity of the geodesic flow was first proved by Hedlund [Hedlund, 1934] for some special finite area groups, and was established in general for the finite area case by Hopf [Hopf, 1936]. The mixing property of the geodesic flow was obtained by Hedlund [Hedlund, 1939], and a simpler proof was given just after this by Hopf [Hopf, 1939]. For the horocyclic flow the mixing property was established by Parasjuk [Parasjuk, 1953]. Quite recently, Marcus [Marcus, 1978] has shown that the horocyclic flow has a much deeper property called mixing of all degrees.

Our presentation combines the ideas of many authors and is simpler and more direct than the original proofs. It includes some elements from Hopf's 1939 treatise [Hopf, 1939] and also a very recent and delightfully elegant approach to the mixing property of the horocyclic flow due to Weissenborn [Weissenborn, 1980].

We start by showing that for a discrete group Γ acting in the unit disk D, a bounded integrable function on Ω which is invariant under the geodesic flow is invariant also under the horocyclic flow. This result is just as easy to prove without the finite area assumption and so it will be stated and proved in this generality. To begin with, we need the following standard result.

Lemma 10.5.1   Suppose f is a bounded, integrable function on Ω, then there exists a sequence $\{f_n\}$ of continuous functions with compact support such that

$$\int_\Omega |f - f_n|\, dM \;\rightarrow\; 0 \;\; as\; n \rightarrow \infty.$$

Theorem 10.5.2   If f is a bounded integrable function on Ω which is invariant under g_t then it is also invariant under h_s.

Proof. Choose s real and we will show that $f(h_s(P)) = f(P)$ for almost all $P \in \Omega$. For the remainder of the proof s remains fixed at this value. Let $\{f_n\}$ be a sequence of functions, continuous of compact support, converging to f as in lemma 10.5.1. For $\epsilon > 0$ let N be such that for any t,

$$\int_\Omega |f(h_{s\,e^{-t}}\, g_t(P)) - f_N(h_{s\,e^{-t}}\, g_t(P))|\, dM = \int_\Omega |f(g_t(P)) - f_N(g_t(P))|\, dM$$

$$= \int_\Omega |f(P) - f_N(P)|\, dM$$

$$< \frac{\epsilon}{3}$$

where we have used the fact that both flows are measure preserving. Now choose t so large that

$$\int_\Omega |f_N(h_{s\,e^{-t}}\, g_t(P)) - f_N(g_t(P))|\, dM \;<\; \frac{\epsilon}{3}\;.$$

Putting these inequalities together we see that, for this choice of t,

$$\int_\Omega |f(h_{s\,e^{-t}}\, g_t(P)) - f(g_t(P))|\, dM \;<\; \epsilon.$$

However, f is invariant under g_t and so

$$f(g_t(P)) = f(P) \;\; and \;\; f(h_{s\,e^{-t}}\, g_t(P)) = f(g_t\, h_s(P)) = f(h_s(P))\;.$$

Thus, for any $\epsilon > 0$,

$$\int_\Omega |f(h_s(P)) - f(P)|\, dM \;<\; \epsilon$$

and so $f(h_s(P)) = f(P)$ almost everywhere as required. $\square$

    Consider now, for a discrete group Γ, the space Ω and let Φ be a function defined on Ω which is invariant under both the horocyclic and the geodesic flows. We may regard Φ as a function on Ω_D which is invariant under Γ and under both flows.

Choose $\xi \in \partial D$ and consider two elements $(u, v, \xi), (u', v', \xi)$ in Ω_D. These two line elements determine geodesics which end at ξ.

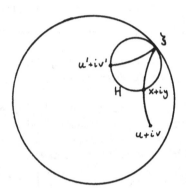

Figure 10.5.1

Let H be the horocycle at ξ through $u' + iv'$, and let $x + iy$ be the point where this horocycle meets the geodesic through $u + iv$. Clearly $\Phi(u', v', \xi) = \Phi(x, y, \xi)$ since Φ is invariant under the horocyclic flow. Also $\Phi(x, y, \xi) = \Phi(u, v, \xi)$ since Φ is invariant under the geodesic flow. Thus we may define a function f on ∂D by

$$f(\xi) = \Phi(u, v, \xi).$$

If Φ is measurable then so is f and, noting that for $\gamma \in \Gamma$

$$\gamma(u, v, \xi) = (Re\ \gamma(u + iv), Im\ \gamma(u + iv), \gamma(\xi))$$

we see that f is invariant under Γ.

As a simple consequence of the definition of ergodicity we have the following result.

Theorem 10.5.3   If Γ is ergodic on ∂D and ϕ is a measurable function on Ω which is invariant under both flows then ϕ is constant almost everywhere.

We can now prove the ergodicity of the geodesic flow.

Theorem 10.5.4   If Γ is a Fuchsian group of finite area then the geodesic flow on Ω is ergodic.

Proof. Let $A \subset \Omega$ be a measurable subset of Ω which has positive measure and is invariant under the geodesic flow. The characteristic function 1_A is integrable and is invariant under the geodesic flow. From theorem 10.5.2 we note that 1_A is also invariant under the horocyclic flow. If we can show that Γ is ergodic on ∂D then we may appeal to theorem 10.5.3 to deduce that 1_A is constant almost everywhere and this in turn would imply that g_t is ergodic. However, the ergodicity of Γ on ∂D is an immediate consequence of theorems 1.6.3 and 6.2.3. $\square$

We next aim to show that the horocyclic flow is ergodic for finite area groups.

Theorem 10.5.5   If Γ is a Fuchsian group of finite area then the horocyclic flow on Ω_D / Γ is ergodic.

Proof. Recall that the flows are related by

$$h_s = R_{\pi - \alpha} \circ g_t \circ R_{2\pi - \alpha}$$

where $\cot \alpha = s/2$ and $s = 2\sinh(t/2)$. If f is invariant under the horocyclic flow then for any $P \in \Omega$,

$$f(P) = f(h_s(P)) = f(R_{\pi-\alpha} \, g_t \, R_{2\pi-\alpha}(P))$$

and so

$$f(R_\alpha \, g_{-t}(P)) = f(R_{\pi-\alpha}(P)) \, .$$

If λ is any bounded integrable function on Ω then

$$\int_\Omega f(R_\alpha \, g_{-t}(P)) \, \lambda(P) \, dM = \int_\Omega f(R_{\pi-\alpha}(P)) \, \lambda(P) \, dM \, .$$

By a change of variables, since g_t is measure preserving,

$$\int_\Omega f(R_\alpha(P)) \, \lambda(g_t(P)) \, dM = \int_\Omega f(R_{\pi-\alpha}(P)) \, \lambda(P) \, dM \, . \qquad (10.5.1)$$

We wish now to consider what happens as $t \to \infty$. We need the following lemma (which we use again later when studying mixing properties). The proof is omitted as it is a standard estimate employing the density of continuous functions of compact support in the space of bounded measurable functions on Ω.

Lemma 10.5.6   Let f be a bounded measurable function on Ω then

$$\lim_{\beta \to 0} \int_\Omega |f(R_\beta(P)) - f(P)| \, dM = 0 \, .$$

Using this lemma in conjunction with equation (10.5.1) and the fact that $\alpha \to 0$ as $t \to \infty$ we have

$$\lim_{t \to \infty} \int_\Omega f(P)\lambda(g_t(P))\, dM \;=\; \int_\Omega f(R_\pi(P))\lambda(P)\, dM \qquad (10.5.2)$$

for any bounded integrable function λ on Ω. From this it follows easily that

$$\lim_{T \to \infty} \frac{1}{T} \int_0^T \left[\int_\Omega f(P)\lambda(g_t(P))\, dM \right] dt \;=\; \int_\Omega f(R_\pi(P))\lambda(P)\, dM \;.$$

By theorem 7.2.4 we can use Fubini's theorem on the left hand side of this equation rewriting it as

$$\lim_{T \to \infty} \int_\Omega \left[\frac{1}{T} \int_0^T \lambda(g_t(P))\, dt \right] f(P)\, dM$$

which, by the bounded convergence theorem, becomes

$$\int_\Omega f(P) \left[\lim_{T \to \infty} \frac{1}{T} \int_0^T \lambda(g_t(P))\, dt \right] dM \;.$$

However, since the geodesic flow is ergodic, we note from theorem 7.2.11 that the term in square brackets above is equal almost everywhere to the constant

$$\frac{1}{M(\Omega)} \int_\Omega \lambda(P)\, dM \;.$$

Thus we have shown that

$$\int_\Omega f(P)\, dM \int_\Omega \lambda(P)\, dM \;=\; M(\Omega) \int_\Omega f(R_\pi(P))\lambda(P)\, dM \qquad (10.5.3)$$

for any bounded integrable function λ and any bounded integrable function f which is invariant under the horocyclic flow. Thus

$$\int_\Omega \lambda(P) \left[f(R_\pi(P)) - \frac{\int_\Omega f(P)\, dM}{M(\Omega)} \right] dM \;=\; 0$$

and since this is true for any bounded integrable λ we have

$$f(R_\pi(P)) \;=\; \frac{\int_\Omega f(P)\, dM}{M(\Omega)}$$

almost everywhere on Ω. Thus f is constant almost everywhere on Ω. We have shown that any bounded, integrable, h_s-invariant function reduces to a constant and the proof is complete. $\square$

Now we go on to consider the mixing property of the geodesic flow. In order to prove mixing it is clearly sufficient to show that for any bounded measurable functions f, λ on Ω

$$\lim_{t \to \infty} \int_\Omega f\,(g_{-t}(P))\,\lambda(P)\,dM \;=\; \frac{\int_\Omega f\,dM \;\int \lambda\,dM}{M(\Omega)}. \qquad (10.5.4)$$

Once again we observe that it is sufficient to prove this result for continuous functions of compact support on Ω. The following lemma is an immediate consequence of the continuity of the horocyclic flow.

Lemma 10.5.7 If f, λ are continuous functions of compact support on Ω then for any $\epsilon > 0$ there exists $\delta > 0$ such that for all t

$$\left| \int_\Omega \left[f\,(g_{-t}(P)) - \frac{1}{\delta} \int_0^\delta f\,(g_{-t}\,h_s(P))\,ds \right] \lambda(P)\,dM \right| < \epsilon.$$

We are now in a position to prove our result.

Theorem 10.5.8 For a Fuchsian group of finite area, the geodesic flow on Ω is mixing.

Proof. Let f, λ be continuous functions of compact support on Ω then for any real t

$$\int_\Omega \left[\frac{1}{\delta} \int_0^\delta f\,(g_{-t}\,h_s(P))\,ds \right] \lambda(P)\,dM \;=\; \int_\Omega \left[\frac{1}{\delta} \int_0^\delta f\,(h_{se^t}\,g_{-t}(P))\,ds \right] \lambda(P)\,dM$$

$$= \int_\Omega \left[\frac{1}{\delta} \int_0^\delta f\,(h_{se^t}(P))\,ds \right] \lambda(g_t(P))\,dM$$

$$= \int_\Omega \left[\frac{1}{\delta e^t} \int_0^{\delta e^t} f\,(h_x(P))\,dx \right] \lambda(g_t(P))\,dM$$

where we have used theorem 10.4.4 and the invariance of the measure M under the geodesic flow. Since the horocycle flow is ergodic we have

$$\lim_{t \to \infty} \frac{1}{\delta e^t} \int_0^{\delta e^t} f\,(h_x(P))\,dx \;=\; \frac{\int_\Omega f\,dM}{M(\Omega)}$$

which we use in the above to obtain

$$\lim_{t \to \infty} \int_\Omega \left[\frac{1}{\delta} \int_0^\delta f\,(g_{-t}\,h_s(P))\,ds \right] \lambda(P)\,dM \;=\; \frac{\int_\Omega f\,dM \;\int \lambda\,dM}{M(\Omega)}.$$

In view of lemma 10.5.5 we have established (10.5.4) for continuous functions of compact support and the theorem is proved . $\square$

For Fuchsian groups of infinite area one would expect the relation (10.5.4) to hold with zero on the right hand side. In fact this is true, and has been proved by the author [Nicholls, 1986]. We prove a slightly weaker version of the result here. In this proof we will have to use an ergodic theorem from Chapter 8.

Theorem 10.5.9   For a Fuchsian group of infinite area which diverges at the exponent 1, the geodesic flow is of zero type. In other words, for f, λ bounded integrable functions on Ω

$$\lim_{t \to \infty} \int_\Omega f\left(g_{-t}(P)\right) \lambda(P)\, dM \;=\; 0 .$$

Proof.   We observe that it is sufficient to prove this result for continuous functions f, λ of compact support. Using lemma 10.5.7 and proceeding exactly as in the proof of theorem 10.5.8 we note that it is sufficient to show that, with $U = \delta e^t$,

$$\int_\Omega \left[\frac{1}{U} \int_0^U f\left(h_u(P)\right) du \right] \lambda(g_t(P))\, dM$$

converges to zero as $t \to \infty$. Now for fixed t we apply the Cauchy-Schwarz inequality, and noting that

$$\int_\Omega \lambda^2(g_t(P))\, dM \;=\; \int_\Omega \lambda^2(P)\, dM$$

is bounded, it remains only to show that $1/U \int_0^U f\left(h_u(P)\right) du$ converges to zero (L^2) as $U \to \infty$. We remark that this quantity is, by theorem 7.2.6, a bounded integrable function invariant under h_s. To complete the proof then, we will show that any bounded integrable h_s-invariant function u is zero. If λ is any bounded integrable function on Ω, we may proceed as in the proof of theorem 10.5.5 to obtain

$$\int_\Omega u\left(R_\pi(P)\right) \lambda(P)\, dM = \int_\Omega u(P) \left[\lim_{T \to \infty} \frac{1}{T} \int \lambda(g_t(P))\, dt \right] dM .$$

The term in braces in the right hand side is, by theorem 7.2.6, an integrable g_t-invariant function on Ω. But, by theorem 8.3.4, g_t is ergodic and so this term must be constant. However, being integrable over a space of infinite measure, this constant must be zero. Thus the right hand side above is zero for any λ, and it follows that the function u is zero as required. $\square$

Now for the horocyclic flow.

Theorem 10.5.10   For a Fuchsian group of finite area the horocyclic flow on Ω is mixing.

Proof. (Weissenborn). For any real t and measurable subsets A, B of Ω

$$M\left(h_t(A) \cap B\right) = M\left[R_{\pi-\alpha}\, g_s\, R_{2\pi-\alpha}(A) \cap B\right]. \qquad (10.5.5)$$

Where we have used theorem 10.3.2 and we have used $\cot \alpha = \sinh(s/2) = t/2$. Since a rotation clearly preserves the measure M we have from (10.5.5)

$$M\left(h_t(A) \cap B\right) = M\left[g_s\, R_{2\pi-\alpha}(A) \cap R_{\pi+\alpha}(B)\right]. \qquad (10.5.6)$$

Now for arbitrary sets X_1, X_2, Y_1, Y_2

$$(X_1 \cap X_2) \nabla (Y_1 \cap Y_2) \subset (X_1 \nabla Y_1) \cup (X_2 \nabla Y_2)$$

where ∇ denotes the symmetric difference. So

$$(g_s(A) \cap R_\pi(B)) \nabla (g_s\, R_{-\alpha}(A) \cap R_{\pi+\alpha}(B)$$

$$\subset (g_s(A) \nabla g_s R_{-\alpha}(A)) \cup (R_\pi(B) \nabla R_{\pi+\alpha}(B))$$

$$= g_s\, (A \nabla R_{-\alpha}(A)) \cup (R_\pi(B) \nabla R_{\pi+\alpha}(B)). \qquad (10.5.7)$$

If we consider for example

$$M(\tilde{A} \cap R_{-\alpha}(A)) = \int_{\tilde{A}} 1_A\, (R_{-\alpha}(P))\, dM$$

then, from lemma 10.5.6, $M(\tilde{A} \cap R_{-\alpha}(A)) \to 0$ as $\alpha \to 0$. From this it follows that the measure of the right hand side of (10.5.7) tends to zero as α tends to zero, i.e., as t tends to infinity. Combining this observation with equation (10.5.6) we have

$$\lim_{t \to \infty} M\left(h_t(A) \cap B\right) = \lim_{t \to \infty} M\left[g_s(A) \cap R_\pi(B)\right].$$

But we know that the geodesic flow is mixing and so

$$\lim_{t \to \infty} M\left(h_t(A) \cap B\right) = \frac{M(A)\, M(R_\pi(B))}{M(\Omega)} = \frac{M(A)\, M(B)}{M(\Omega)}$$

and the proof is complete. $\square$

10.6 Unique Ergodicity

In this section we investigate a phenomenon which is peculiar to the horocyclic flow. Since we have not been concerned with this topic for the major part of the book we have deferred its definition to this point. Let us suppose we have a flow f_s on a space Ω then f_s is said to be **uniquely ergodic** if there exists a unique Borel probability measure on Ω which is invariant under f_s. Let us denote this measure by μ and suppose $A \subset \Omega$ is measurable and f_s-invariant. If $\mu(A) > 0$ we may form a new measure M on Ω given by

$$M(E) = \frac{\mu(E \cap A)}{\mu(A)}$$

for any Borel subset E of Ω. The reader will check immediately that M is a Borel probability measure on Ω which is invariant under f_s. Thus M must coincide with μ and we are forced to the conclusion that $\mu(A) = 1$. We have shown that f_s is ergodic. But more than this is true. Knowing that f_s is ergodic and that $\mu(\Omega) = 1$ we may appeal to theorem 7.2.8 to obtain that, for any $\lambda \in L^1(\Omega)$,

$$\lim_{T \to \infty} \frac{1}{T} \int_0^T \lambda(f_s(x)) ds = \int_\Omega \lambda \, d\mu \tag{10.6.1}$$

almost everywhere in Ω. Now suppose A is a set of positive μ measure, then we may take $\lambda = 1_A$ in the above and obtain a limit almost everywhere. But suppose that for some $x \in \Omega$ (10.6.1) is false. In this case we may find a sequence $T_n \to \infty$ with

$$\lim_{n \to \infty} \frac{1}{T_n} \int_0^{T_n} 1_A(f_s(x)) ds = C \neq \mu(A). \tag{10.6.2}$$

Now we define a measure M_n in the following way

$$M_n(E) = \frac{1}{T_n} \int_0^{T_n} 1_E(f_s(x)) ds$$

It is routine to check that M_n is a Borel probability measure. In the topology of weak convergence some subsequence M_{n_k} converges to a Borel probability measure on Ω. Call this new measure M and note that for any real u

$$M(f_u(E)) = \lim_{n \to \infty} \frac{1}{T_n} \int_0^{T_n} 1_{f_u(E)}(f_s(x)) ds$$

$$= \lim_{n \to \infty} \frac{1}{T_n} \int_0^{T_n} 1_E(f_{s-u}(x)) ds$$

$$= \lim_{n \to \infty} \frac{1}{T_n} \int_{-u}^{T_n - u} 1_E(f_s(x)) ds = \lim_{n \to \infty} \frac{1}{T_n} \int_0^{T_n} 1_E f_s(x) ds$$

$$= M(E).$$

M is thus flow invariant and must therefore agree with μ. However, (10.6.2) guarantees that $M(A) \neq \mu(A)$ and we have a contradiction. Thus, for indicator functions, (10.6.1) must be true **everywhere** in Ω. But it follows easily from this that (10.6.1) is true everywhere in Ω for **any** L^1 function λ.

In the other direction, suppose μ is a Borel probability measure on Ω which is invariant under f_s and that for any $\lambda \in L^1(\Omega)$

$$\lim_{T \to \infty} \frac{1}{T} \int_0^T \lambda(f_s(x)) ds = \int_\Omega \lambda d\mu$$

for all $x \in \Omega$. If M is another Borel probability measure invariant under f_s then we may appeal to theorem 7.2.6, applied to the indicator function of any Borel set, to see that $M \equiv \mu$.

We have established the following result.

Theorem 10.6.1   The flow f_s on Ω is uniquely ergodic if and only if there exists a Borel probability measure μ on Ω, invariant under f_s and with

$$\lim_{T \to \infty} \frac{1}{T} \int_0^T \lambda(f_s(x)) ds = \int_\Omega \lambda d\mu$$

for every L^1 function λ and every $x \in \Omega$. In this case, μ is the unique measure guaranteed by unique ergodicity.

We define now the notion of minimality. The flow f_s on Ω is said to be **minimal** if every trajectory is dense.

Theorem 10.6.2   If the flow f_s is uniquely ergodic then it is minimal.

Proof. Let μ be the unique Borel probability measure on Ω invariant under f_s. If the flow is not minimal then there exists $x \in \Omega$ and an open subset A of Ω with $f_1(x) \cap A = \varnothing$ for all s. We note that

$$\lim_{T \to \infty} \frac{1}{T} \int_0^T 1_A(f_s(x)) ds = 0 \neq \mu(A)$$

and by theorem 10.6.1, the flow is not uniquely ergodic. $\square$

We now consider the geodesic and horocyclic flows for Fuchsian groups.

Theorem 10.6.3   If Γ is a non-elementary Fuchsian group then the geodesic flow is not minimal.

Proof. Since the group is non-elementary it contains a hyperbolic element, γ say. If γ fixes ξ and η then, for any s real, the trajectory of the element (ξ, η, s) is a closed loop on the quotient space. $\square$

The horocyclic flow may be minimal. Our next result was first proved by Hedlund [Hedlund, 1936].

Theorem 10.6.4   If Γ is a Fuchsian group of finite area then the horocyclic flow is minimal if and only if there are no parabolic elements in Γ.

Proof. If Γ contains a parabolic element then it is well known that a horocycle H may be erected at the fixed point which is invariant under the stabilizer and mapped outside itself by every other element of the group. Any line element determining H does not have a dense trajectory. For the converse, suppose Γ contains no parabolic elements. Every point of ∂B is a conical limit point for Γ and we deduce from lemma 2.5.2 that every horocycle has images of Euclidean radius arbitrarily close to 1. We now appeal to Hedlund's argument [Hedlund, 1936 p. 537] to see that any horocycle has Γ-images approximating any other horocycle. In other words, every trajectory under the horocycle flow is dense. $\square$

We conclude this chapter by proving unique ergodicity of the horocycle flow. The result was first established by Furstenberg [Furstenberg, 1973]. Our proof follows the method due to Marcus [Marcus, 1975] which can be used even in the case of two-dimensional manifolds of variable negative curvature. In much of the literature relating to unique ergodicity, the horocyclic flow is defined somewhat differently than the way we have defined it in this chapter. Consequently, we will introduce a new flow ϕ_s on Ω — this is the version used in [Marcus, 1975] for example — and show how it is related to "our" horocycle flow h_s.

Given $x \in \Omega$ denote by H_x the horocycle through the base point of x which is tangent to the unit circle at the initial point of the directed geodesic determined by x. For s real denote by $\psi_s(x)$ the line element based on H_x obtained by sliding x around H_x a counterclockwise directed distance s. It should be clear that, with D_α denoting a directed rotation through α on Ω, $\phi_s \equiv D_\pi h_s D_\pi$. It is immediate from this that h_s is uniquely ergodic if and only if ϕ_s has that property. Thus we will concentrate on proving that ϕ_s is uniquely ergodic. It is routine to check that, with g_t denoting the geodesic flow,

$$g_t \circ \phi_s = \psi_{se^t} \circ g_t. \tag{10.6.3}$$

Start by defining a sequence $t_n = n \log 2$ (so $e^{t_n} = 2^n$) and a sequence $\{R_n\}$ of operators on continuous functions on Ω given by

$$R_n f(x) = \frac{1}{2^n} \int_0^{2^n} f \circ \phi_s \circ g_{t_n}(x) ds.$$

Our aim will be to show that $\{R_n f(x)\}$ converges uniformly to a constant as $n \to \infty$. If this is the case then with f given and $\epsilon > 0$ chosen we find n so that

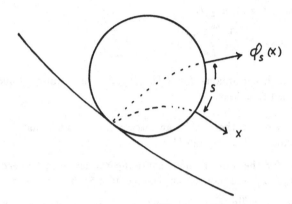

$$\phi_s(x)$$

Figure 10.6.1

$$\lceil R_n f(x) - C \rceil < \epsilon$$

for all $x \in \Omega$. Thus

$$R_n f(g_{-t_n} \phi_{2^n i}(x)) = \frac{1}{2^n} \int_0^{2^n} f \circ \phi_{s + 2^n i}(x) ds$$

$$= \frac{1}{2^n} \int_{2^n i}^{2^n (1+i)} f \circ \phi_s(x) ds$$

and the average of the sequence $\{R_n f(g_{-t_n} \circ \phi_{2^n i}(x))\}_{i=0}^{j-1}$ is

$$\frac{1}{2^n j} \int_0^{2^n j} f \circ \phi_s(x) ds.$$

Thus, for t sufficiently large

$$\left| \frac{1}{t} \int_0^t f \circ \phi_s(x) ds - c \right| < 2\epsilon$$

and, for all $x \in \Omega$, $\lim \frac{1}{t} \int_0^t f \circ \phi_s(x) ds = c$. From theorem 7.2.8, the constant c must be equal to $\int_\Omega f d\mu$ and, by theorem 10.6.1., we have shown that the horocycle flow is uniquely ergodic.

Our problem thus reduces to showing that $R_n f(x)$ converges uniformly to a constant as $n \to \infty$. We need three lemmas. We state these lemmas with the assumption in each that R_n is the sequence of operators defined above and that the underlying Fuchsian group is of finite area with no parabolic elements.

Lemma 10.6.5   For each n and $m \geq 0$

$$R_{n+m} f = \frac{1}{2^m} \sum_{j=0}^{2^m-1} R_n f \circ \phi_j \circ g_{t_m} .$$

Lemma 10.6.6   For each continuous f on Ω, $\{R_n f\}$ is an equi-continuous family of uniformly bounded functions.

Lemma 10.6.7   For any $\epsilon > 0$ there exists an integer $N > 0$ such that for all $y \in \Omega$, $\{\phi_k(y) : k = 0,1,2,...,N\}$ is ϵ dense in Ω.

We defer the proofs for the moment and, assuming the lemmas, proceed as follows. Set $c_n = \min_{x \in \Omega} R_n f(x)$ and note from lemma 10.6.5 that c_n is a non decreasing sequence. Set $c = \lim c_n$. Now let $\{n_k\}$ be a sequence with $R_{n_i} f \to F$ uniformly (by lemma 10.6.6). Note that F is continuous with minimum c. By lemma 10.6.5

$$R_{n_i+m} f \to \frac{1}{2^m} \sum_{j=0}^{2^m-1} F \circ \phi_j \circ g_{t_m} \equiv \overline{F}.$$

Now $\overline{F}$ also has minimum c. If $\overline{F}(x_0) = c$ then $F(\phi_j \circ g_{t_m}(x_0)) = c$ for $0 \leq j \leq 2^m - 1$. But since m is arbitrary we see that F takes the values c on an ϵ dense set (lemma 10.6.7). Thus $F \equiv c$ and, by the Arzela-Ascoli theorem, $R_n f \to c$ uniformly.

Thus, modulo the three lemmas above, we have.

Theorem 10.6.8   If Γ is a Fuchsian group of finite area and with no parabolic elements then the horocyclic flow is uniquely ergodic.

Proof of lemma 10.6.5 By definition,

$$R_n f(x) = \frac{1}{2^n} \int_0^{2^n} f \circ \phi_s \circ g_{t_n}(x) ds$$

$$= \frac{1}{2^n} \int_0^{2^n} f \circ g_{t_n} \circ \phi_{se^{-t_n}}(x) ds$$

$$= \frac{1}{2^n} \int_0^{2^n} f \circ g_{t_n} \circ \phi_{s/2^n}(x) ds$$

$$= \int_0^1 f \circ g_{t_n} \circ \phi_s(x) ds \qquad (10.6.4)$$

where we have used the definition of t_n, and the equation (10.6.3). Thus using these same things over again we will have

$$R_n f \circ \phi_j \circ g_{t_m} = \int_0^1 f \circ g_{t_n} \circ \phi_{s+j} \circ g_{t_m}(x) ds$$

$$= \int_j^{j+1} f \circ g_{t_n} \circ \phi_s \circ g_{t_m}(x) ds$$

$$= \int_j^{j+1} f \circ \phi_{se^{t_n}} \circ g_{t_{n+m}}(x) ds$$

$$= \frac{1}{2^n} \int_{2^n j}^{2^n(j+1)} f \circ \phi_s \circ g_{t_{n+m}}(x) ds.$$

It follows that

$$\frac{1}{2^n} \sum_{j=0}^{2^n-1} R_n f \circ \phi_j \circ g_{t_m} = \frac{1}{2^{m+n}} \sum_{j=0}^{2^m-1} \int_{2^n j}^{2^n(j+1)} f \circ \phi_s \circ g_{t_{n+m}}(x) ds$$

$$= \frac{1}{2^{m+n}} \int_0^{2^{m+n}} f \circ \phi_s \circ g_{t_{n+m}}(x) ds$$

$$= R_{n+m} f$$

and this completes the proof of the lemma. $\square$

Proof of lemma 10.6.6. Suppose y, $x \in \Omega$, let H_z be the horocycle through the base point of z that is internally tangent to the unit circle at the initial point of the geodesic determined by z. Figure 10.6.2 illustrates the situation. The geodesic from this point of tangency to the end point of the geodesic determined by y yields a unique element of Ω, denoted $[y,z]$, such that $H_{[y,z]} = H_z$. From the definition we see that, for some u real,

$$\phi_u(z) = [y,z].$$

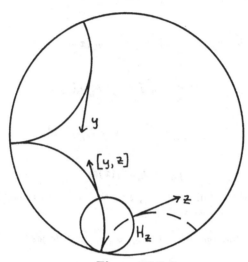

Figure 10.6.2

Now for $x, y \in \Omega$ and s real define a function $k_{xy}(s)$ by

$$\phi_{k_{xy}(s)}(x) = [\phi_s(y), x].$$

This is shown in figure 10.6.3.

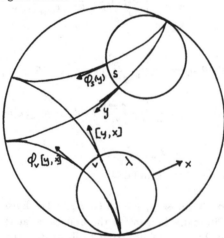

Figure 10.6.3

We need to gather some information on the function k. Refer to figure 10.6.3 where we have assumed $s > 0$. Note that λ is a function of x, y only and that

$$k_{xy}(s) = v + \lambda.$$

With x,y fixed, $k_{xy}(s)$ is an increasing function of s. We need information on the behavior of $k_{xy}(s)$ if x is close to y. In order to obtain this we conjugate to the upper half-plane to obtain figure 10.6.4 in which t and r are the Euclidean radii of the two horocycles.

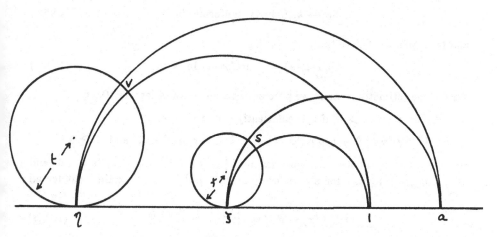

Figure 10.6.4

The transform

$$V(z) = \frac{(a+1-\xi)(z-1)+1-\xi}{z-\xi}$$

preserves the upper half-plane, maps ξ to ∞ and fixes 1 and a. It is easily checked that the horocycle at ξ is mapped to the horocycle

$$\text{Im } z = \frac{(1-\xi)(a-\xi)}{2r}$$

from which it follows trivially that

$$s = \frac{a-1}{(1-\xi)(a-\xi)} \cdot 2r.$$

A similar calculation for v shows that

$$\frac{v}{s} = \left(\frac{t}{r}\right)\left(\frac{1-\xi}{1-\eta}\right)\left(\frac{a-\xi}{a-\eta}\right)$$

which gives v as a function of s. Note that t, r, ξ, η are determined by the line elements x, y whereas a is also dependent upon s. It is evident from this relationship however that, given $\epsilon > 0$ there exists $\delta > 0$ such that for all $s, t -1 \leq s < t \leq 1$ and x, y satisfying $d(x,y) < \delta$

$$\left| \frac{k_{xy}(t) - k_{xy}(s)}{t - s} - 1 \right| < \epsilon \tag{10.6.5}$$

This shows that k_{xy} is Lipschitz and hence that

$$\lim_{y \to x} \frac{d}{ds} k_{xy}(s) = 1 \text{ uniformly in } s \tag{10.6.6}$$

and from this we deduce that

$$\lim_{y \to x} k_{xy}(s) = s \text{ uniformly in } s . \tag{10.6.7}$$

With these estimates in hand we proceed with the proof of lemma 10.6.6.

Given x and $\epsilon > 0$, if y is sufficiently close to x then

$$|f \circ g_t(\phi_s(y)) - f \circ g_t([\phi_s(y),x])| < \epsilon \text{ for } |s| < 1 \text{ and } t \geq 0.$$

By our working above, we may assume $|k_{xy}(s) - s| < \epsilon$ for $|s| < 1$ and $|(d/ds)k_{xy}(s) - 1| < \epsilon$ for almost every s with $|s| < 1$. We claim that for all $n \geq 0$

$$|R_n f(x) - R_n f(y)| < \epsilon \cdot (1 + 3\|f\|). \tag{10.6.8}$$

To see this, note that by choice of x, y, and ϵ

$$\left| R_n f(y) - \int_0^1 f \circ g_{t_n}([\phi_s(y),x]) ds \right|$$

$$= \left| \int_0^1 (f \circ g_{t_n} \circ \phi_s(y) - f \circ g_{t_n}([\phi_s(y),x])) ds \right| < \epsilon.$$

Now by definition of k_{xy}

$$\int_0^1 f \circ g_{t_n}([\phi_s(y),x]) ds = \int_0^1 f \circ g_{t_n} \phi_{k_{xy}(s)}(x) ds.$$

Since $|(d/ds)k_{xy}(s) - 1| < \epsilon$ we have

$$\left| \int_0^1 f \circ g_{t_n} \circ \phi_{k_{xy}(s)}(x) ds - \int_0^1 (\frac{d}{ds} k_{xy}(s)) \cdot (f \circ g_{t_n} \circ \phi_{k_{xy}(s)}(x)) ds \right| < \epsilon \|f\|$$

Since k is Lipschitz, hence absolutely continuous, we can use the change of variables formula

$$\int_0^1 (\frac{d}{ds} k_{xy}(s)) \cdot (f \circ g_{t_n} \circ \phi_{k_{xy}(s)}(x)) ds = \int_{k_{xy}(0)}^{k_{xy}(1)} f \circ g_{t_n} \circ \phi_s(x) ds$$

and deduce that

$$\left| \int_{k_{xy}(0)}^{k_{xy}(1)} f \circ g_{t_n} \circ \phi_s(x) ds - R_n f(x) \right| < \| f \| |k_{xy}(1) - 1 + k_{xy}(0)| < 2\epsilon \| f \| .$$

Now the last four displayed formulae yield (10.6.8) and the proof of the lemma is complete. □

Proof of lemma 10.6.7 This result follows immediately from the compactness of Ω if we can prove that the map ϕ_1 is minimal. By a suitable reparametrization of the flow it is sufficient to show that for some positive real t the map ϕ_t is minimal. Let us suppose to the contrary that every map ϕ_t is non-minimal. Choose $x_0 \in \Omega$ and, for each t define

$$A_t = \text{closure } \{\phi_{nt}(x_0) : n \in Z\} .$$

We note that A_t is a closed, ϕ_t-invariant proper subset of Ω, and that ϕ_t restricted to A_t is minimal. Given t, the set of reals $s > 0$ such that $A_s = A_t$ must comprise rational multiples of t and must be bounded away from zero, otherwise a dense subset of the trajectory $\{\phi_u(x_0) : u > 0\}$ lies in A_t, and this contradicts the fact that the trajectory is dense in Ω (theorem 10.6.4). Select a minimal such s and define a function f_s on the trajectory of x_0 by

$$f(\phi_u(x_0)) = e^{2\pi i u/s} f(x_0), \quad f(x_0) = 1 .$$

If we have $y \in \Omega$ with $\phi_{u_n}(x_0) \to y$ and $\phi_{v_n}(x_0) \to y$ then $\phi_{u_n - v_n}(x_0) \to x_0$ and so, by our remarks above, the fractional part of $(u_n - v_n)/s$ tends to zero. Thus the function f extends to a continuous eigenfunction for the flow; i.e., a function f satisfying $f(\phi_u(x)) = e^{2\pi i u/s} f(x)$, $|f| = 1$. There are clearly uncountably many different eigenfunctions, and this contradicts the separability of the space of continuous functions on Ω. □

10.7 A Lattice Point Problem

For a Fuchsian group Γ we recall the orbital counting function

$$N(r, x, y) = \text{card } \{\gamma \in \Gamma : \rho(x, \gamma(y)) < r\} .$$

We have by now established several estimates for this counting function, and we are concerned in this section with its asymptotic behavior as $r \to \infty$. This can be viewed as the hyperbolic analog of the Gauss circle problem. We will derive asymptotic formulae both in the finite area and in the infinite area case and will also consider angular distribution of orbits. All the results we give here will generalize to the n-dimensional setting, the proofs given here will extend — [Nicholls, 1983b], [Nicholls, 1986] — although the definition of the horocyclic flow in n-dimensions leads to certain complications. The results we give are due to S.J. Patterson [Patterson, 1975], and [Patterson, 1977] although his proofs are quite different from ours, using the Selberg theory of the spectral decomposition of the Laplace operator. Similar analytic methods have been employed by Lax and Phillips [Lax and Phillips, 1982] to obtain asymptotic estimates of the counting function for a wide class of discrete groups of Euclidean and non-Euclidean motions.

We assume throughout this section that Γ is a Fuchsian group preserving the unit disc and that no element of Γ (except the identity) fixes the origin. Thus the region D_0 defined in section 1.4 is a fundamental domain for Γ. If D_0 has finite hyperbolic area then we say that Γ is a group of finite area, and we write $V(\Gamma) = V(D_0)$.

We will be concerned with estimates on the proportion of circles centered at the origin which are covered by group images of some fixed disc. Let Δ be a hyperbolic disc of center a and radius δ, we will assume that a is not fixed by any element of Γ (except the identity) and that δ is chosen so small that Γ-images of Δ are disjoint. We write w for angular measure and define

$$g(s) \;=\; w(\{x : \rho(x,0) = s\} \cap \Gamma(\Delta)) \;.$$

We have the following result.

Lemma 10.7.1   The function $g(s)$ is uniformly continuous on $(0,\infty)$.

Proof. Define $n(s,\Delta)$ to be the number of $\gamma \in \Gamma$ with the property that $\gamma(\Delta) \cap \{x : \rho(x,0) = s\} \neq \varnothing$. An upper bound on $n(s,\Delta)$ is easily obtained via an area argument — in fact this is what we did in theorem 1.5.1 — and we have

$$n(s,\Delta) < k \; V\{x : \rho(x,0) < s\} \qquad s \geq s_0 \qquad (10.7.1)$$

where k is an absolute constant. Now consider a single image Δ' of Δ and we claim that the difference in angular measures between the sets $\Delta' \cap \{x : \rho(x,0) = s - \epsilon\}$ and $\Delta' \cap \{x : \rho(x,0) = s\}$ is maximized if Δ' is internally tangent to the circle $\{x : \rho(x,0) = s\}$. To see this, note that for s large and ϵ small (compared to δ), the intersection $\Delta' \cap \{x : \rho(x,0) = s\}$ is essentially a chord of Δ', and the difference above reduces to a multiple of the difference in angle subtended at the Euclidean center of Δ' by two parallel chords. Our claim follows easily from this.

For s large and δ small we illustrate the situation in figure 10.7.1 and we calculate that λ is asymptotic (as $s \to \infty$) to $\epsilon \delta e^{-s}$. Thus the difference in angular measures between the sets $\Delta' \cap \{x : \rho(x,0) = s - \epsilon\}$ and $\Delta' \cap \{x : \rho(x,0) = s\}$ is bounded by a quantity which, as $s \to \infty$, is asymptotic to a constant times $\delta \epsilon e^{-s}$. Assuming this maximum difference is attained for every single image Δ' of Δ which intersects the disc $\{x : \rho(x,0) \leq s\}$ we obtain

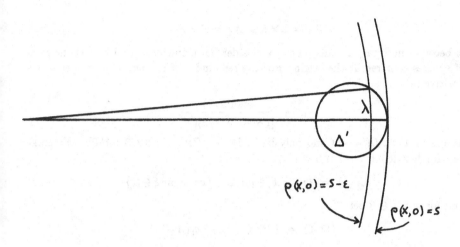

Figure 10.7.1

from the above, and (10.7.1), that $|g(s) - g(s-\epsilon)|$ is bounded above by a quantity that is asymptotic (as $s \to \infty$) to

$$k\, \delta\epsilon\, e^{-s}\, V\{x : \rho(x,0) \leq s\}.$$

The proof of the lemma is complete when we note that $V\{x : \rho(x,0) \leq s\}$ is asymptotic to a constant times e^s. $\square$

If Γ is a Fuchsian group of finite area then the geodesic flow is mixing. Interpreting this property in the ball we have the following result.

Lemma 10.7.2   If Γ is a Fuchsian group of finite area and if A_1, A_2 are measurable subsets of $D_0 \times S$ then

$$\lim_{t \to \infty} M\Big[\Gamma(A_1) \cap g_t(A_2)\Big] = \frac{M(A_1)M(A_2)}{2\pi V(D_0)}.$$

From this we prove our main covering lemma.

Lemma 10.7.3   If Γ is a Fuchsian group of finite area and if $\Delta \subset D_0$ is a hyperbolic disc then

$$\lim_{t \to \infty} \frac{w\Big[\{x : \rho(x,0) = t\} \cap \Gamma(\Delta)\Big]}{2\pi} = \frac{V(\Delta)}{V(D_0)}.$$

Proof. Let $a \in D_0$ be fixed and let Δ be a fixed hyperbolic disc centered at a. If C is a ball centered at the origin of hyperbolic radius r we define

$$A_1 = \Delta \times S , \quad A_2 = C \times S .$$

It becomes necessary at this point to consider both Euclidean and hyperbolic radii of spheres centered at the origin. For the remainder of the proof, whenever s is a positive real we set

$$s_1 = \frac{\sinh(s/2)}{[1 + \sinh^2(s/2)]^{1/2}}$$

noting that $\rho(x,0) = s$ if and only if $|x| = s_1$ [Beardon, 1983 p.130]. We make two further definitions. If $0 < t < \infty$ then set

$$X(t) = \{x : (x,\xi) \in g_t(A_2) \text{ for some } \xi \in S\}$$

and for any $x \in B$ set

$$I(x,t) = \{\xi \in S : (x,\xi) \in g_t(A_2)\} .$$

Trivially, $I(x,t) \neq \emptyset$ if and only if $x \in X(t)$. It follows from the symmetry of the situation that if $\rho(x_1,0) = \rho(x_2,0) = s$ then

$$w[I(x_1,t)] = w[I(x_2,t)] = L(s,t)$$

say. We are now in a position to use lemma 10.7.2, and we observe that

$$M[\Gamma(A_1) \cap g_t(A_2)] = \int_{X(t)} w[I(x,t)] 1_{\Gamma(\Delta)}(x) dV(x)$$

$$= \int_{t-r}^{t+r} \frac{L(s,t) g(s) s_1}{(1 - s_1^2)^2} \cdot \frac{ds_1}{ds} ds$$

where we recall that $X(t)$ is the annulus $\{x : t - r < \rho(x,0) < t + r\}$. Now observe that $M(A_2) = M(g_t(A_2))$ and

$$M(g_t(A_2)) = \int_{X(t)} w[I(x,t)] dV(x) = \int_{t-r}^{t+r} \frac{L(s,t) w(S) s_1}{(1 - s_1^2)^2} \cdot \frac{ds_1}{ds} ds .$$

We thus write $M[\Gamma(A_1) \cap g_t(A_2)]/M(A_2)$ as the quotient of two integrals and, from the continuity of the integrands, we see that for some s satisfying $t - r < s < t + r$ we have $M[\Gamma(A_1) \cap g_t(A_2)]/M(A_2) = g(s)/w(S)$. However, by lemma 10.7.1, g is uniformly continuous and so, given $\epsilon > 0$, we find $r > 0$ so small that

$$\left| \frac{M[\Gamma(A_1) \cap g_t(A_2)]}{M(A_2)} - \frac{g(t)}{w(S)} \right| < \epsilon .$$

From lemma 10.7.2,

$$\frac{M[\Gamma(A_1) \cap g_t(A_2)]}{M(A_2)} \rightarrow \frac{M(A_1)}{2\pi V(D_0)}$$

as $t \rightarrow \infty$. We note that $M(A_1) = 2\pi V(\Delta)$ and deduce that

$$\lim_{t \rightarrow \infty} \frac{g(t)}{2\pi} = \frac{V(\Delta)}{V(D_0)}$$

as required. $\square$

From the covering lemma we obtain an asymptotic estimate for the orbital counting function.

Theorem 10.7.4   Let Γ be a Fuchsian group of finite area. If x_1, $x_2 \in B$ then

$$N(s,x_1,x_2) \sim \frac{V\{x : \rho(x,0) < s\}}{V(\Gamma)} \qquad \text{as } s \rightarrow \infty .$$

Proof. Upon integration we obtain from lemma 10.7.3

$$\lim_{t \rightarrow \infty} \frac{V[\{x : \rho(x,0) < t\} \cap \Gamma(\Delta)]}{V\{x : \rho(x,0) < t\}} = \frac{V(\Delta)}{V(D_0)} . \qquad (10.7.2)$$

Now choose $\epsilon > 0$ and find $\delta > 0$ so small that for $t > t_0$ say,

$$\frac{V\{x : \rho(x,0) < t+\delta\}}{V\{x : \rho(x,0) < t\}} < 1+\epsilon ,$$

which may be done since $V\{x : \rho(x,0) < t\} \sim ke^t$. We apply (10.7.2) to the ball Δ of radius δ and deduce that, for $t > t_1$ say,

$$\frac{V[\{x : \rho(x,0) < t+\delta\} \cap \Gamma(\Delta)]}{V\{x : \rho(x,0) < t+\delta\}} < \frac{V(\Delta)}{V(D_0)} \cdot (1+\epsilon) .$$

Now if $\rho(\gamma(a),0) < t$ then $\gamma(\Delta) \subset \{x : \rho(x,0) < t+\delta\}$, and so

$$V(\Delta) N(t,0,a) \leq V[\{x : \rho(x,0) < t+\delta\} \cap \Gamma(\Delta)] .$$

Using the inequalities above we see that, for $t > t_1$,

$$\frac{N(t,0,a)}{V\{x : \rho(x,0) < t\}} \leq \frac{(1+\epsilon)^2}{V(\Gamma)} .$$

A similar lower bound shows that

$$N(t,0,a) \sim \frac{V\{x : \rho(x,0) < t\}}{V(\Gamma)} \qquad (10.7.3)$$

as $t \rightarrow \infty$.

Now consider two points x_1, x_2 in B and let γ be a Moebius transform with $\gamma(x_1) = 0$, $\gamma(x_2) = w$, say. We write $G = \gamma\Gamma\gamma^{-1}$ and note that $V(G) = V(\Gamma)$. Clearly,

$$\rho(x_1, g(x_2)) = \rho(\gamma(x_1), \gamma(g(x_2))) = \rho(0, \gamma g \gamma^{-1}(w))$$

for any $g \in \Gamma$. It follows that $N_\Gamma(s, x_1, x_2) = N_G(s, 0, w)$, and the theorem follows from (10.7.3). $\square$

For Fuchsian groups of infinite area we have the following result of Patterson [Patterson, 1977].

Theorem 10.7.5 If Γ is a Fuchsian group of infinite area, and if x_1, $x_2 \in B$ then

$$N(s, x_1, x_2) = o\left(V\{x : \rho(x, 0) < s\}\right) \quad \text{as} \quad s \to \infty.$$

Proof. If Γ diverges at the exponent 1 then the proof will follow along exactly the same lines as that of theorem 10.7.4 with the role of the mixing property of the geodesic flow being played by the zero type property (theorem 10.5.9). In the case that Γ converges at the exponent 1 we have

$$\sum_{\gamma \in \Gamma} e^{-\rho(0, \gamma z)} < \infty.$$

Thus the integral $\int_1^\infty e^{-s} dN(s, 0, x)$, and hence also the integral $\int_1^\infty e^{-s} N(s, 0, x) ds$ converge. It is now clear that

$$N(s, 0, x) = o(e^s) \quad \text{as} \quad s \to \infty$$

and so

$$N(s, 0, x) = o\left(V\{x : \rho(x, 0) < s\}\right),$$

and the general result follows by conjugation as in the proof of theorem 10.7.4. $\square$

Our last result concerns angular distribution of orbits. If θ is an interval of the unit circle, s is positive, and $a \in B$ then define $n(\theta, a, s)$ to be the number of $\gamma \in \Gamma$ with $\rho(0, \gamma(a)) < s$ and $\arg \gamma(a) \in \theta$.

Theorem 10.7.6 If Γ is a Fuchsian group of finite area then

$$n(\theta, a, s) \sim \frac{w(\theta)}{2\pi} \cdot \frac{V\{x : \rho(x, 0) < s\}}{V(\Gamma)} \quad \text{as} \quad s \to \infty.$$

Proof. We need a covering lemma analogous to lemma 10.7.3. If we define Θ to be the cone subtended at the origin by θ then we have the following.

Lemma 10.7.7   If Γ is a Fuchsian group of finite area and if $\Delta \subset D_0$ is a hyperbolic disc then

$$\lim_{t \to \infty} \frac{w[\{x : \rho(x,0) = t\} \cap \Theta \cap \Gamma(\Delta)]}{w(\theta)} = \frac{V(\Delta)}{V(D_0)} .$$

The proof of theorem 10.7.6 follows from this lemma in the same way that theorem 10.7.4 follows from lemma 10.7.3. Thus it remains only to prove the lemma.

We fix Δ and define $A_1 = \Delta \times S$, now let C be a disc centered at the origin and of hyperbolic radius r. We define $A_2 = C \times \theta$. It is useful to observe that

$$\Theta(t) = \Theta \cap \{x : \rho(x,0) < t\} = \bigcup_{0 < s < t} g_t(\{0\} \times \theta)$$

and so $g_t(A_2)$ is (for small r) a thin shell whose cross-section is approximately $g_t(\{0\} \times \theta)$ (i.e., $\Theta \cap \{x : \rho(x,0) = t\}$), together with a set of directions at each point.

Explicitly, this shell is given by $X(t) = \{x : (x,\xi) \in g_t(A_2) \text{ for some } \xi \in S\}$ and the set of directions at each point x of $X(t)$ is given by

$$I(x,t) = \{\xi \in S : (x,\xi) \in g_t(A_2)\} .$$

For economy of notation we have used the same symbols $X(t)$ and $I(x,t)$, as were used in the proof of lemma 10.7.3 — their meaning of course is different now. The shell $X(t)$ is a subset of the annular region $\{x : t - r < \rho(x,0) < t + r\}$. It is important to note that, contrary to the previous situation, the values of the angular measure w of two direction sets $I(x_1,t)$, $I(x_2,t)$ associated with points x_1, x_2 of equal modulus in $X(t)$ are not necessarily equal. This fact gives rise to an added difficulty in computing $M[\Gamma(A_1) \cap g_t(A_2)]/M(A_2)$, which is required for the application of lemma 10.7.2.

We will show that given t and s satisfying $t - r < s < t + r$, then

$$X(t) \cap \{x : \rho(x,0) = s\}$$

comprises an *admissible part*, any two points of which have direction sets of the same angular magnitude, and an *inadmissible part* which, for r close enough to zero, is so small as to make no difference to our asymptotic estimates.

Accordingly, suppose $x \in X(t)$. Then, for some $\xi \in S$, $(x,\xi) \in g_t(A_2)$ and so the sphere $\{y : \rho(x,y) = t\}$ intersects C. Join each point z of this intersection to x by a geodesic ray and let ξ_z be the direction at z which determines this geodesic. We say that x is *admissible* if each such ξ_z belongs to θ. To put it another way, if $x \in X(t)$ then x is obtained by moving a distance t from a point of C along a geodesic in a direction belonging to θ. The point x is *admissible* if it can be so obtained from *any* point of C which is distant t from x. Figure 10.7.2 illustrates an admissible point x and the set of directions $I(x,t)$.

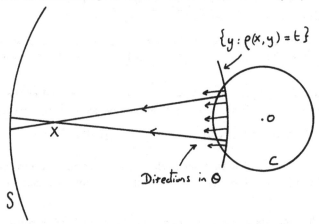

Figure 10.7.2

It will be seen that if x_1, $x_2 \in X(t)$ are both admissible and if $|x_1| = |x_2|$, then $I(x_2,t)$ is merely a rotation of $I(x_1,t)$ and consequently the two sets have the same angular measure.

As regards the inadmissible set we have the following.

Lemma 10.7.8   Given $\epsilon > 0$ there exists $r_0 > 0$ such that if $r < r_0$ and s, t satisfy $1 < t - r < s < t + r$ then   the   inadmissible   part   of $X(t) \cap \{x : \rho(x,0) = s\}$ has angular measure less than ϵ.

Proof. Suppose $x \in X(t)$ is inadmissible. Then there exists a point $z \in C$ with $\rho(z,x) = t$ and such that the geodesic connecting z to x determines a direction at z which does not belong to θ. On the other hand, x is obtained by moving a distance t from a point of C along a geodesic in a direction belonging to θ. We may as well suppose that this latter point is on the radius joining 0 to x. Figure 10.7.3 illustrates the situation.

     A straightforward calculation shows that the Euclidean separation of the two points ξ_1, ξ_2 of figure 10.7.3 is $O(r)$ as $r \to 0$ provided that $\rho(x,0) > 1$, say. It follows that the radial projection of x onto S has a separation from the boundary of θ which is of the order $O(r)$. Thus the radial projection of the inadmissible set onto S is contained in a band of width $O(r)$ around the edge of θ. Since θ is an interval, this band has an area which approaches zero with r.

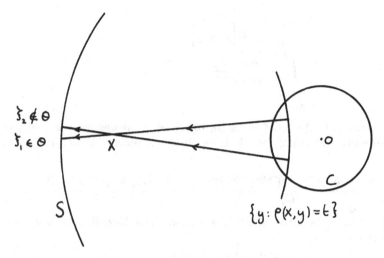

Figure 10.7.3

Since the angular measure of our inadmissible set is the area of its projection onto S, we see that the lemma is proved. □

We return to the proof of lemma 10.7.7 and note that

$$M\left[\Gamma(A_1) \cap g_t(A_2)\right] = \int_{X(t)} w\left[I(x,t)\right] 1_{\Gamma(\Delta)}(x) dV(x). \qquad (10.7.4)$$

The right side of (10.7.4) is computed as a double integral — first over the sphere $\{x : \rho(x,0) = s\}$, and then radially letting s vary from $t-r$ to $t+r$. We approximate this first integral, replacing $X(t) \cap \{x : \rho(x,0) = s\}$ by the slightly smaller set $\Theta \cap \{x : \rho(x,0) = s\}$ and then assuming that every point in the latter set is admissible. The error obtained may be estimated using lemma 10.7.8. For admissible x in $X(t)$ satisfying $\rho(x,0) = s$, we denote by $L(s,t)$ the angular measure $w\left[I(x,t)\right]$.

Given $\epsilon > 0$ we find r_0 so small that if $r < r_0$ and $t > 1$, say, then

$$\frac{M\left[\Gamma(A_1) \cap g_t(A_2)\right]}{\int_{t-r}^{t+r} \frac{L(s,t)s_1}{(1-s_1^2)^2} w\left[\{x : \rho(x,0) = s\} \cap \Theta \cap \Gamma(\Delta)\right] \frac{ds_1}{ds} ds}$$

differs from 1 by at most ϵ.

Similarly, for the same values of r and t,

$$\frac{M\left[g_t(A_2)\right]}{\displaystyle\int_{t-r}^{t+r} \frac{L(s,t)s_1}{(1-s_1^2)^2}\, w(\theta)\, \frac{ds_1}{ds}\, ds}$$

differs from 1 by at most ϵ.

We proceed as in the proof of lemma 10.7.3 using the uniform continuity of $w[\{x : \rho(x,0) = s\} \cap \Theta \cap \Gamma(\Delta)]$ (which follows from the proof of lemma 10.7.1), to deduce that

$$\lim_{t \to \infty} \frac{w[\{x : \rho(x,0) = t\} \cap \Theta \cap \Gamma(\Delta)]}{w(\theta)} = \lim_{t \to \infty} \frac{M[(\Gamma(A_1) \cap g_t(A_2)]}{M[A_2]}.$$

By lemma 10.7.2, this latter limit is equal to $M(A_1)/(2\pi V(D_0))$ and since $M(A_1) = 2\pi V(\Delta)$, we see that

$$\lim_{t \to \infty} \frac{w[\{x : \rho(x,0) = t\} \cap \Theta \cap \Gamma(\Delta)]}{w(\theta)} = \frac{V(\Delta)}{V(D_0)}.$$

This completes the proof of the theorem. □

References

Agard, 1982.
> Stephen Agard, *Elementary Properties of Moebius Transformations in R^n with Applications to Rigidity Theory,* Mathematics Report 82-110, School of Mathematics, University of Minnesota, Minneapolis, 1982.

Agard, 1983.
> Stephen Agard, "A geometric proof of Mostow's rigidity theorem for groups of divergence type," *Acta Math.* **151**, pp. 231-252, 1983.

Ahlfors, 1981.
> L.V. Ahlfors, *Moebius transformations in several dimensions,* Lecture Notes, School of Mathematics, University of Minnesota, Minneapolis, 1981.

Ahlfors and Sario, 1960.
> L.V. Ahlfors and L. Sario, *Riemann Surfaces,* Princeton University Press, Princeton 1960.

Apanasov, 1982.
> B.N. Apanosov, "Geometrically finite groups of transformations of space," *Siberian Math. J.* **23**, pp. 771-780, 1982.

Apanasov, 1983.
> B.N. Apanasov, "Geometrically finite hyperbolic structures on manifolds," *Ann. Glob. Analysis & Geometry* **1**, pp. 1-22, 1983.

Artin, 1924.
> E. Artin, "Ein mechanische system mit quasiergodischen bahnen," *Abh. Math. Sem. Univ. Hamburg* **3**, pp. 170-175, 1924.

Beardon, 1968.
> A.F. Beardon, "The exponent of convergence of Poincaré series," *Proc. London Math. Soc.* **(3) 18**, pp. 461-483, 1968.

Beardon, 1983
> A.F. Beardon, *The geometry of discrete groups,* Springer Verlag, New York 1983.

Beardon and Maskit, 1974.
> A.F. Beardon and B. Maskit, "Limit points of Kleinian groups and finite sided fundamental polyhedra," *Acta Math.* **132**, pp. 1-12, 1974.

Beardon and Nicholls, 1982.
    A.F. Beardon and P.J. Nicholls, "Ford and Dirichlet regions for
    Fuchsian groups," *Canadian J. Math.* **34**, pp.806-815, 1982.

Billingsley, 1986.
    P. Billingsley, *Probability and Measure, 2nd* Edition, John Wiley
    and Sons, New York 1986.

Bowditch, 1988.
    B. H. Bowditch, *Geometrical Finiteness for Hyperbolic Groups,*
    Mathematics Institute, University of Warwick, Coventry, 1988.

Davis and Morgan, 1984.
    M.W. Davis and J.W. Morgan, "Finite group actions on
    homotopy 3-spheres," in *The Smith Conjecture,* Academic Press,
    Orlando, Florida, 1984.

Federer, 1969.
    H. Federer, *Geometric Measure Theory,* Springer Verlag, New
    York 1969.

Fomin and Gel'fand, 1952.
    S.V. Fomin and I.M. Gel'fand, "Geodesic flows on manifolds of
    constant negative curvature," *Uspehi Mat. Nauk* **7**, pp. 118-137,
    1952.

Furstenberg, 1973.
    H. Furstenberg, "The unique ergodicity of the horocycle flow," in
    Recent Advances in Topological Dynamics, Springer-Verlag, New
    York, 1973.

Gottschalk and Hedlund, 1955.
    W.H. Gottschalk and G.A. Hedlund, *Topological Dynamics,*
    American Mathematical Society, Providence 1955.

Greenberg, 1962.
    L. Greenberg, "Discrete subgroups of the Lorentz group," *Math.
    Scand* **10**, pp. 85-107, 1962.

Greenberg, 1966.
    L. Greenberg, "Fundamental polyhedra for Kleinian groups,"
    Annals of Math. **84**, pp. 433-441, 1966.

Hedlund, 1935.
    G.A. Hedlund, "A metrically transitive group defined by the
    modular group," *American Jour. Math.* **57**, pp. 668-678, 1935.

Hedlund, 1936.
    G.A. Hedlund, "Fuchsian groups and transitive horocycles," *Duke
    Math. J.* **2**, pp. 530-542, 1936.

Hedlund, 1939.
> G.A. Hedlund, "Fuchsian groups and mixtures," *Annals of Math.*
> **40**, pp. 370-383, 1939.

Hopf, 1936.
> E. Hopf, "Fuchsian groups and ergodic theory," *Transactions American Math. Soc.* **39**, pp. 299-314, 1936.

Hopf, 1939.
> E. Hopf, "Statistik der geodatischen linien in mannigfaltigkeiten negativer krummung," *Ber. Verh. Sachs. Akad. Wiss. Leipzig* **91**, pp. 261-304, 1939.

Koebe, 1930.
> P. Koebe, "Riemannsche mannigfaltigkeiten und nicht euklidische raumformen VI," *S.-B. Deutsche. Akad. Wiss. Berlin K1. Math. Phys. Tech.* , pp. 504-541, 1930.

Lax and Phillips, 1982.
> P.D. Lax and R.S Phillips, "The asymptotic distribution of lattice points in Euclidean and non-Euclidean spaces," *J. Funct. Anal.* **46**, pp. 280-350, 1982.

Lehner, 1964.
> J. Lehner, *Discontinuous groups and automorphic functions,* American Mathematical Society, Providence 1964.

Lobell, 1929.
> F. Lobell, "Uber die geodatischen linien der Clifford - Kleinschen flachen," *Math. Zeit.* **30**, pp. 572-607, 1929.

Marcus, 1975.
> D. Marcus, "Unique ergodicity of the horocycle flow : variable negative curvature case," *Israel Journal of Math.* **21**, pp. 133-144, 1975.

Marcus, 1978.
> B. Marcus, "The horocycle flow is mixing of all degrees," *Inventiones Math.* **46**, pp. 201-209, 1978.

Marden, 1974.
> A. Marden, "The geometry of finitely generated Kleinian groups," *Ann. Math.* **99**, pp. 383-462, 1974.

Myrberg, 1931.
> P.J. Myrberg, "Ein approximationsatz fur die Fuchssen gruppen," *Acta Math.* **57**, pp. 389-409, 1931.

Nicholls, 1980.
> P.J. Nicholls, "Garnett points for Fuchsian groups," *Bull. London Math. Soc.* **12**, pp. 216-218, 1980.

Nicholls, 1981a.
>   P.J. Nicholls, "The boundary behavior of automorphic forms,"
>   *Duke Math. J.* **48**, pp. 807-812, 1981.

Nicholls, 1981b.
>   P.J. Nicholls, "Kleinian groups of divergence type," *Proc. Amer.*
>   *Math. Soc.* **83**, pp. 319-324, 1981.

Nicholls, 1983a.
>   P.J. Nicholls, "Discrete groups on the sphere at infinity," *Bull.*
>   *London Math. Soc.* **15**, pp. 488-492, 1983.

Nicholls, 1983b.
>   P.J. Nicholls, "A lattice point problem in hyperbolic space,"
>   *Michigan Math. J.* **30**, pp. 273-287, 1983.

Nicholls, 1984.
>   P.J. Nicholls, "Ford and Dirichlet regions for discrete groups of
>   hyperbolic motions," *Trans. Amer. Math. Soc.* **282**, pp. 355-365,
>   1984.

Nicholls, 1986.
>   P.J. Nicholls, "The geodesic flow for discrete groups of infinite
>   volume," *Proc. Amer. Math. Soc.* **96**, pp. 311-317, 1986.

Parasjuk, 1953.
>   O. Parasjuk, "Flows of horocycles on surfaces of constant negative
>   curvature (in Russian)," *Uspehi Mat. Nauk* **8**, pp. 125-126, 1953.

Patterson, 1975.
>   S.J. Patterson, "A lattice point problem in hyperbolic space,"
>   *Mathematika* **22**, pp. 81-88, 1975.

Patterson, 1976a.
>   S.J. Patterson, "The limit set of a Fuchsian group," *Acta*
>   *Math.* **136**, pp. 241-273, 1976.

Patterson, 1976b.
>   S.J. Patterson, "The exponent of convergence of Poincaré series,"
>   *Monat. fur* Math. **82**, pp. 297-315, 1976.

Patterson, 1977.
>   S.J. Patterson, "Spectral theory and Fuchsian groups," *Math.*
>   *Proc. Cambs. Philos. Soc.* **81**, pp. 59-75, 1977.

Patterson, 1987.
>   S.J. Patterson, "Lectures on measures on limit sets of Kleinian
>   groups," in *Analytical and geometric aspects of hyperbolic space,*
>   London Math. Society Lecture Notes 111, pp. 281-323, Cambridge
>   Univ. Press, 1987.

Poincaré, 1883.
> H. Poincaré, "Memoire sur les groupes Kleineens," *Acta Math.* **3**, pp. 49-92, 1883.

Pommerenke, 1976.
> Ch. Pommerenke, "On the Green's function of Fuchsian groups," *Ann. Acad. Sci. Fenn. A1 Math.* **2**, pp. 409-427, 1976.

Rodin and Sario, 1968.
> B. Rodin and L. Sario, *Principal Functions*, Van Nostrand, Princeton 1968.

Rudolph, 1982.
> D.J. Rudolph, "Ergodic behaviour of Sullivan's geometric measure on a geometrically finite hyperbolic manifold," *Ergod. Th. and Dynam. Sys.* **2**, pp. 491-512, 1982.

Sario and Nakai, 1970.
> L. Sario and M. Nakai, *Classification theory of Riemann surfaces*, Springer Verlag, New York 1970.

Satake, 1956.
> I. Satake, "On a generalization of the notion of manifold," *Proc. Nat. Acad. Sci. USA.* **42**, pp. 359-363, 1956.

Seidel, 1935.
> W. Seidel, "On a metric property of Fuchsian groups," *Proc. Nat. Acad. Sci. USA* **21**, pp. 475-478, 1935.

Selberg, 1960.
> A. Selberg, *On discontinuous groups in higher-dimensional symmetric spaces*, Contributions to function theory, Tata Institute, Dombay, pp. 147-164, 1960.

Sheingorn, 1980a.
> M. Sheingorn, "Transitivity for the modular group," *Math. Proc. Camb. Phil. Soc.* **88**, pp. 409-423, 1980.

Sheingorn, 1980b.
> M. Sheingorn, "Boundary behavior of automorphic forms and transitivity for the modular group," *Illinois J. Math.* **24**, pp.440-451, 1980.

Shimada, 1960.
> S. Shimada, "On P.J. Myrberg's approximation theorem on Fuchsian groups," *Mem. Coll. Sci. Kyoto U. Ser. A.* **33**, pp. 231-241, 1960.

Sprindzuk, 1979.
> V.G. Sprindzuk, *Metric theory of Diophantine approximation*, Wiley, New York, 1979.

Sullivan, 1979.
> D. Sullivan, "The density at infinity of a discrete group of hyperbolic motions," *Publ. Math. I.H.E.S.* **50**, pp. 171-202, 1979.

Sullivan, 1981.
> D. Sullivan, *On the ergodic theory at infinity of an arbitrary discrete group of hyperbolic motions*, Ann. of Math. Studies, 97, pp. 465-496, 1981.

Sullivan, 1982.
> D. Sullivan, "Discrete conformal groups and measurable dynamics," *Bull. Amer. Math. Soc.* **6** pp. 57-73, 1982.

Sullivan, 1984.
> D. Sullivan, "Entropy, Hausdorff measures old and new, and limit sets of geometrically finite Kleinian groups," *Acta Math.* **153** pp. 259-277, 1984.

Titchmarsh, 1939.
> E.C. Titchmarsh, *The theory of functions, 2nd Ed.* Oxford University Press, Oxford 1939.

Tsuji, 1959.
> M. Tsuji, *Potential theory in modern function theory*, Maruzen, Tokyo, 1959.

Tukia, 1984.
> P. Tukia, "The Hausdorff dimension of the limit set of a geometrically finite Kleinian group," *Acta Math.* **152** pp. 127-140, 1984.

Tukia, 1985.
> P. Tukia, "Differentiability and rigidity of Moebius groups," *Invent. Math.* **82** pp. 557-578, 1985.

Tuller, 1938.
> A. Tuller, "The measure of transitive geodesics on certain three dimensional manifolds, "*Duke Math. J.* **4**, pp. 78-94, 1938.

Weissenborn, 1980.
> G. Weissenborn, "Ergodische eigenschaften Fuchsser gruppen," Dissertation, Technische Universitat Berlin, 1980.

Wolf, 1974.
> J. A. Wolf, *Spaces of constant curvature, 4th ed.*, Publish or Perish, Berkeley, 1977.

INDEX OF SYMBOLS

INDEX